Historic Industrial Scenes

THE MINES OF SHROPSHIRE

Preesgweene Colliery, 1873. This colliery, near Weston Rhyn on the Oswestry Coalfield, had a number of shafts and eventually reached a depth of about 500ft. It was, as can be seen from the number of employees in the photograph, quite a substantial undertaking, but eventually closed in 1891. One surprising feature is the large number of obviously very young children employed shown seated on the ground at the front of the picture. The principal coal seams worked were the Top and Lower Yard, Two Feet, Five Feet and Six Feet Seams. *(Salop County Library)*

Historic Industrial Scenes

THE MINES OF SHROPSHIRE

Ivor J. Brown

Moorland Publishing Company

ISBN 0 903485 32 X

Typeset by
NEWSTEAD PUBLISHING LTD, Nottingham
Printed in Great Britain by
WOOD MITCHELL & CO LTD, Stoke on Trent
for the publishers
MOORLAND PUBLISHING COMPANY

Contents

Introduction	6
The Coalbrookdale Coalfield	8
Other Shropshire Coalfields	
Forest of Wyre Coalfield	68
Clee Hills Coalfields	78
Shrewsbury Coalfield	80
Oswestry Coalfield	82
The South-West Shropshire Metal Mines	83
Bibliography	109
Index	110

Preface

These old photographs turned up in many different and some most unexpected places. One thing was common, however, and that was the thrill of identifying a long lost view of Shropshire's industrial past. How many more old pictures lie hidden in drawers and cupboards it is impossible to guess, but it is certain that once these examples are seen others will be remembered and brought to light. Both the writer and the Ironbridge Gorge Museum Trust Ltd would like very much to know of any such find and perhaps a second volume can be prepared with an even clearer, more representative selection of pictures. It is certain that long forgotten names will spring to mind as faces are recognised and incidents will be recalled. The writer would like to know of these too.

As so much has depended on human memory it is quite possible that some of the detail may not be as accurate as that obtained purely from documentary sources. Every care has been taken, however, to cross-check the information given, but if anything is seen to be wrong the writer takes full responsibility.

The writer hopes that these pictures will give as much pleasure to others as they have given to him. Life and places are changing fast and only by studying contemporary scenes it is possible to understand fully the way of life of past generations. This way of life is being lost as a new town with new industries is formed within the county, but fortunately a Museum Trust has been set up with ideals which will ensure that some features will remain to give meaning to the pictures given here.

The Ironbridge Gorge Museum Trust Ltd., a private charity, was established in 1968 to preserve the industrial heritage of the Coalbrookdale Coalfield and mining plays a large part in this story. Wherever possible relics are preserved on site but the very situation of mining remains, either deep underground or above dangerous mine shafts, usually precludes this possibility. Therefore many items have been taken to the Blists Hill Open Air Museum where they can be seen alongside reconstructed original remains, such as the Blists Hill Pit and steam winder, and replica mining items such as the drift mine. It is hoped that those who read this book will visit the museum, as only by a comparison of the physical remains with the documentary evidence can true understanding be achieved.

The grid reference number of most of the sites is given to enable the exact location to be identified on the Ordnance Survey maps.

Introduction

Shropshire is rich in the variety of minerals available. These range from the ores of the metals lead, zinc, copper and iron, to the fuels coal, oil and peat. A large number of general industrial and building materials are also available, including 'red' brick and tile clays, fireclays, limestone, barytes, pyrites, witherite, calcite, salt, sand and sandstone. Igneous and metamorphic rocks suitable for use as aggregates are also available and are still being extensively quarried. This publication however is concerned only with the underground extraction of minerals and while many historic quarrying photographs exist they are not included in this work.

Five coalfields are found within the county limits. The Coalbrookdale Coalfield is by far the most important, historically and economically. The Forest of Wyre Coalfield, the small and isolated Clee Hills Coalfields and the Shrewsbury Coalfield are all interesting, but have contained very few large or medium-sized mines. Finally, around Oswestry, part of the Denbighshire Coalfield runs into the county, although it has always been more closely associated with industry in North Wales rather than the English Midlands.

The South-West Shropshire metalliferous mining area includes the important recent workings for barytes on the fringes of the metal mining area proper. Elsewhere in the county isolated mining areas exist for copper (Llanymynech, Clive and Eardiston for example) and for limestone including areas around the Wrekin, Lilleshall and, in the south of the country, Novers and Gorstly Rough.

Acknowledgements

The author would like to thank the following persons and organisations for their assistance in the production of this work: S. B. Smith, Ironbridge, for reading Section One and for making many useful suggestions; S. W. Davies for assistance with Section Two; K. C. Lock, Shrewsbury, for reading Section Three and for providing a large number of photographs of the metalliferous mining area; to the following for supplying information: E. V. Edwards, T. H. Stonehouse, the Shropshire County Archivist, the Shropshire Mining Club, the National Coal Board, the Ironbridge Gorge Museum Trust Ltd. (abbreviated to IGMT); and to all who have kindly supplied illustrations, acknowledgement for which is made beneath their respective contributions. Photographs not otherwise acknowledged are from the author's collection.

A special word of thanks is due to my wife for typing and checking and for encouragement throughout the preparation of this publication.

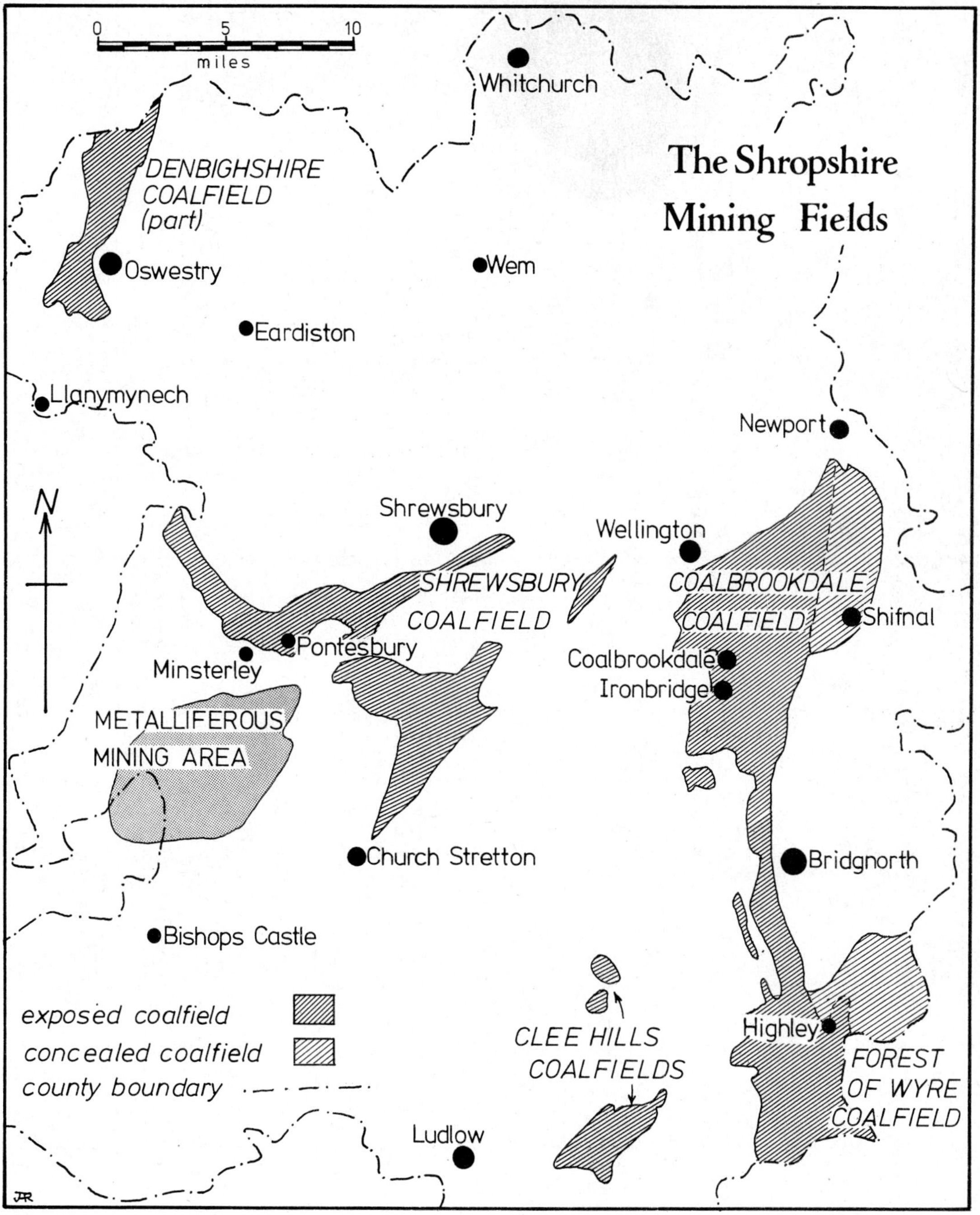
The Shropshire Mining Fields
0
5
10
miles
Whitchurch
DENBIGHSHIRE COALFIELD (part)
Oswestry
Wem
Eardiston
Llanymynech
Newport
N
Shrewsbury
Wellington
SHREWSBURY COALFIELD
COALBROOKDALE COALFIELD
Shifnal
Pontesbury
Minsterley
Coalbrookdale
Ironbridge
METALLIFEROUS MINING AREA
Church Stretton
Bridgnorth
Bishops Castle
exposed coalfield
concealed coalfield
county boundary
CLEE HILLS COALFIELDS
Highley
FOREST OF WYRE COALFIELD
Ludlow
JR

The Coalbrookdale Coalfield

This coalfield has been the most productive of the Shropshire fields. There is evidence which indicates that coal and ironstone were being worked throughout the Middle Ages and coal was certainly being extracted by the Romans long before this. It was however the successful substitution of coke for charcoal in the blast furnace by Abraham Darby in 1709 that led to the great expansion in mineral working in the area and ultimately to Coalbrookdale being dubbed the 'cradle of the Industrial Revolution.' It is often overlooked however that, prior to Darby's innovation, the prime mineral in the area had been ironstone and this mineral remained important until the end of the nineteenth century.

The minerals of the coalfield are found in the following geological series:

Upper Coal Measures	containing high quality red clays but poor coal seams
Middle Coal Measures Lower Coal Measures	containing all the valuable seams of ironstone, coal and fireclay
Carboniferous Limestone	containing the limestones worked at The Hatch, Steeraways and Lilleshall
Silurian Limestone	lying unconformably beneath the Lower Coal Measures and mined in the area of Lincoln Hill

For the miner and geologist the coalfield has posed many problems. These include numerous faults, folds, unconformities and an undulating topography producing many seam outcrops. Other problems required special machinery and techniques, such as mine drainage, the gassy nature of the seams, their general thinness (average about 3ft) and their variation in dip.

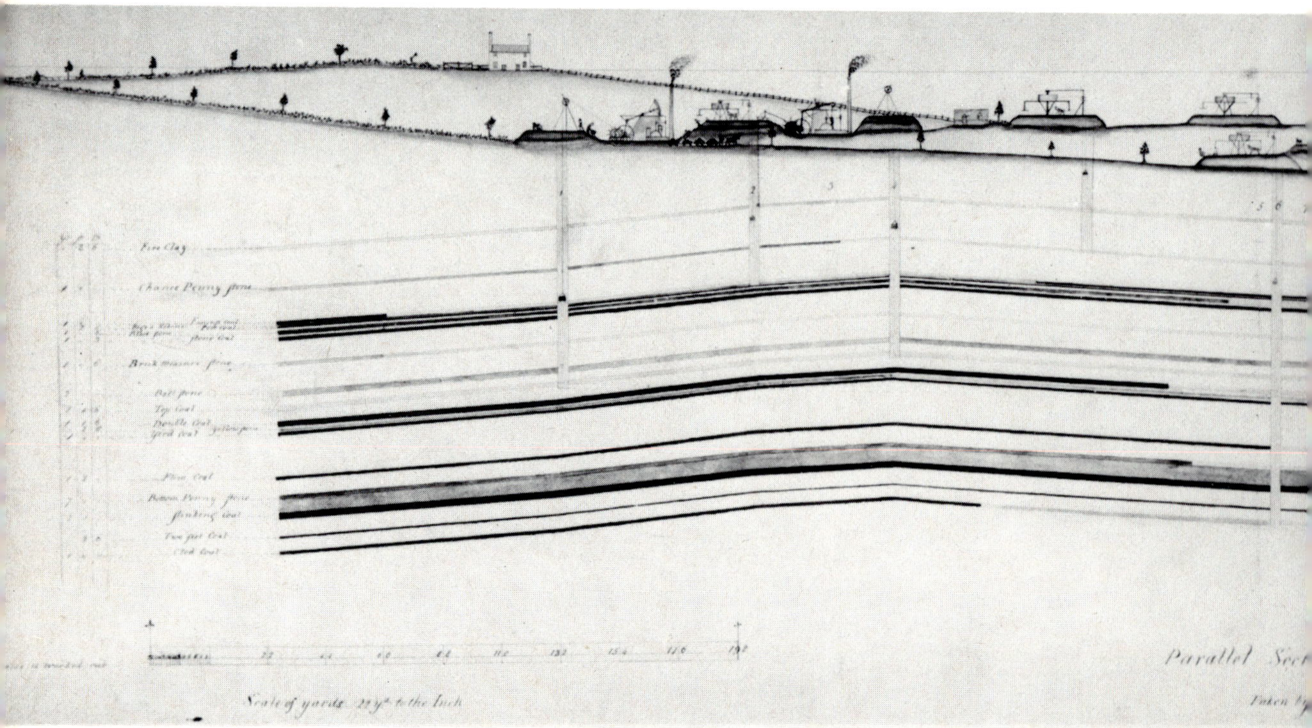

The proven coalfield is roughly rectangular in shape, about 7 miles by 3½ miles and about 24 square miles in area. The recorded peak of coal production was in 1871 when over one million tons of coal was raised. Ironstone production reached its peak two years later, in 1873, when over 430,000 tons was produced. At the time of the Nationalisation of the coal industry in 1947 there were sixteen mines at work in the coalfield, but in 1975 there is only one mine still in operation, the National Coal Board's Granville Colliery.

As a result of the extensive industrialisation of the area in the eighteenth and nineteenth centuries and its subsequent demise, there is now much derelict land. However many of these scars are being actively removed as land is reclaimed for the construction of the new town of Telford which is planned to include most of the coalfield.

1 **'Parallel Section of Hadley Colliery'** by Charles W. Pearce, 1812. This is the earliest illustrated geological section of a mine in the coalfield. It shows that some ten seams of coal, seven of ironstone and one of fireclay were being worked at this time in strata aggregating 300ft thickness. The shafts were up to 400ft deep and the mine equipment included seven horse gins, four whimsey engines and five steam pumping engines. In addition three hand capstan-type winders were in use, probably to assist with shaft repairs. It has been estimated that by 1800 there were over 200 steam engines at work in the coalfield, certainly more per square mile than in any other part of Britain at that time. *(Apley Estates)*

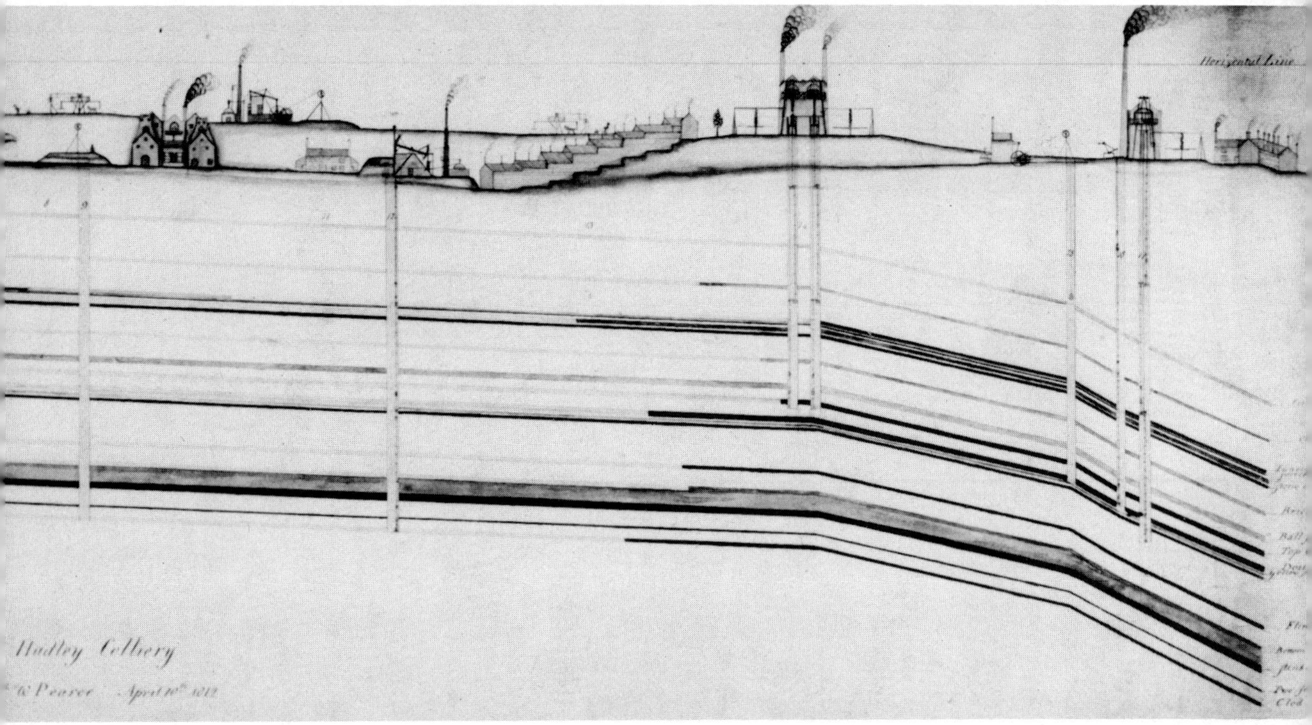

2 **'A view of the mouth of a coal pit near Broseley in Shropshire.'** The print was published in 1788, the artist being George Robertson (c 1748 - 88) and the engraver Francis Chesham. The gin (or winding equipment) seems to be powered by two horses which would indicate that the mine was fairly deep. This is also borne out by the fact that a surface-furnace ventilating system was in use. The chimney from the furnace is shown on the right of the picture. *(IGMT)*

3 **'Lateral mine shafts besides the Severn.'** The drawing was published by Dibden in 1801. At this time Dibden produced an interesting account of the Tar Tunnel and it is possible that one of these 'lateral mine shafts', or adits, represents the mouth of this tunnel. The picture shows a similar tunnel on the opposite river bank but there is no geological or historical evidence for such a tunnel opposite the Tar Tunnel at Coalport. It is more likely therefore that the caverns shown were in fact limekilns at Lincoln Hill and Benthall respectively and that the pyramid of light above the tunnel represents the flames from the kiln. (Also a view of the limekilns at Coalbrookdale by J. M. W. Turner published in 1825 has a similar lighting effect.) *(Shropshire Star and Morley Tonkin Collection)*

4 **Engraving of a specimen of the fossil tree 'Sigillaria'** as discovered in a quarry at Coalbrookdale in 1863. Numerous similar specimens have been found in the Big Flint Sandstone of the area. This sandstone forms the roof of the Pennystone Ironstone Seam, the most productive ironstone measure in the coalfield. Similar specimens have also been found in the Little Flint Sandstone at the base of the Productive Measures. *(The Illustrated London News, 7 Feb 1863)*

5 **Seeking coal at the Black Rock, Coalbrookdale,** about 1900. It is probable that the miners were prospecting for the Little Flint Coal Seam, since the 'Black Rock' is the local name for the sandstone roof of the seam, the name being derived from its heavy impregnation with bitumen. *(D. Oakley)*

6 *(Overleaf, top)* **Footridge at Foxholes Wood, Benthall Edge,** about 1916. (SJ 667031) The 'Footridge' is the local name for an adit mine. The photograph was taken on the occasion of a visit from the Lord of the Manor of Benthall, Lord Forester, who was both land and mineral owner. The equipment shown includes a sledge, with loop for attaching the belt for manual hauling (called 'mobbying'), a riddle, a shovel and a sack. *(The late W. Harper)*

7 *(Overleaf, bottom)* Another view probably taken on the same occasion. In the background miners can be seen at work digging into the coal seam outcrops to fill the sacks. *(D. Oakley)*

FOOTRIDGE
FOX HOLES

8 **A Newcomen-type steam winding engine at Lightmoor** about 1860. The engine is believed to have been made by the Coalbrookdale Company for its own mine, the Gravel Leasows Colliery (SJ 679055). Of particular interest is the spool-type winding drum used with a winding chain. The chain consisted of three separate square-link chains placed side-by-side and held together by wooden pegs (see Fig 82). *(IGMT)*

9 **Old beam engine at 'The Rats Pits' Oakengates** about 1899. This engine stood beside the Great Western Railway between Wellington and Oakengates and was obviously derelict at the time the photograph was taken. The engine has been identified as an early nineteenth-century rotative beam-engine, double-acting, geared for winding and possibly for pumping also. The photograph and description appeared in *Trans Shrops Arch Soc,* Vol LIII, 1949. *(The late L. C. Lloyd)*

THE MADELEY WOOD COMPANY

The Madeley Wood Company was one of the few large multi-industry based companies operating in the area. The company operated mines of coal, ironstone, fireclay, red clay and limestone as well as blast furnaces, brickworks and tileworks. The company owed its origins to Richard Reynolds who was manager of the Coalbrookdale Company from 1763 to 1768, and at the same time ran his own smaller business at nearby Madeley Wood. The Madeley Wood Company prospered throughout the nineteenth century, but after 1900 closure followed closure, culminating in the blowing out of the Blists Hill Blast Furnaces in 1912. During the 1920s the company sold its remaining mine, the Kemberton Mine, and its widespread land and mineral holdings.

10 **Kemberton Mine** (SJ 112055) in 1864. This was the last deep mine sunk by the Madeley Wood Company, it was 364yd deep and was completed in 1864, the date of the picture. The drawing is taken from a geological section of the shaft by William Ward of Madeley. It shows a typical outside winding drum or spool, with the weight of the winding chain supported by 'rolley' posts. The headframe behind the chimney is over a shallower shaft sunk for obtaining 'house' or boiler-feed water.

11 *(opposite top)* **Surface employees at Kemberton** about 1900. A typical line-up of mine workers and horses in front of the shafts. In the middle distance can be seen the engine house, unchanged since 1864, while in the foreground to the left is an unusual view of the squat ventilating furnace chimney. At this time the mine used over sixty horses and ponies underground. *(F. Turner)*

12 *(opposite, bottom)* **View of headframe** in 1945. This photograph shows the substantial tandem wooden headframe constructed after 1912 following the expansion of the mine. Tandem headframes are a feature of the coalfield. Following expansion and concentration on coalgetting the mine employed about 600 persons which, after Nationalisation in 1947, eventually increased to over 800. Some ironstone was obtained however during World War II by hand picking the waste rock as it left the mine.

13 *(below)* **Winding engine houses** 1964. The photo shows the survival of the earliest engine house as offices. To the left of this, off the picture, is the later horizontal steam winding-engine house, in use 1912 to 1937, and behind it the house containing the electric winder used from 1937. This was the first coal mine in Shropshire to have an electric winder. The mine closed in 1967. *(NCB)*

Kemberton Mine underground 1946. This set of three photographs shows the mine immediately before Nationalisation.

14 A typical loading point, where conveyors from the coal face load into wooden wagons on the main haulageway. An endless rope was used to draw the wagons or tubs to the pit bottom.

15 A line of tubs waiting their turn to be loaded onto the mine cage at the pit bottom. The tunnel to the left connected the Kemberton Mine to Halesfield Mine.

16 A Samson electric coal cutter used for under-cutting the coal seam so that it could be blasted down easily. This machine was actually in use in a low development tunnel being used to form a new coal face. After blasting the coal was loaded by hand onto the conveyor to the left of the photograph.

17 The underground locomotive-haulage road, about one mile long and constructed in the early 1950s to supercede the meandering rope haulage road from the working area to pit bottom. The photo was taken about 1958. *(NCB)*

DRE438

Halesfield Mine (SJ 704051) This mine was situated about half a mile from Kemberton Mine to which it was joined underground after 1939. Both mines were owned by the Madeley Wood Company. There were two pairs of shafts sunk between 1840 and 1870 to a depth of about 300yd. For many years the two pairs of shafts were worked independently, despite their depth, and old miners used to relate how they would send messages to their friends in the next pit by striking the solid rock with their picks. The mines were mainly ironstone producers and in 1895 employed a total of 160 men and women, the latter picking the ironstone nodules from the weathered shale on the surface. In 1912 mineral production ceased following the closure of the company's Blists Hill Furnaces, although some coal was produced during the period 1920 to 1925. After 1939 the shafts were used solely for pumping from the joined mines, as upcast air shafts and as an emergency means of egress from the Kemberton Mine.

18 *(opposite, top)* General view of the surface 1946. Shows the two headframes at the mine surrounded by the mine buildings. In the foreground to the left are the Lancashire boilers used for steam raising, to the right the 'new' fan house with the evasee chimney in front. The ropes for both shafts were operated from one winding drum, as one wound off the other wound on. Because of the great distance from the 'outside' pit to the winder an auxiliary pulley can be seen fitted to the inner pit headframe.

19 *(opposite, bottom)* The horizontal steam-engine house and inner pit headframe. The pit is enclosed as it was used as an upcast shaft, the air being drawn up by the fan.

20 *(below)* General view of surface 1963. A similar view to that shown in Fig 18. By this time the steam engine had been replaced by an electric winder and the old wooden headframe by one of steel. In the foreground can be seen the concrete covers over the two shafts of the Old Halesfield or Tom Rose Pits. When working under the chartermaster system two of the shafts worked coal and two ironstone. Tom Rose, after whom the shafts were named, was one of the last chartermasters at the mine.

Blists Hill Mine (SJ 694034) This mine was operated by the Madeley Wood Company until 1916, after which date it was sold to a succession of private operators. It finally closed in 1940. The shafts were almost certainly sunk in the 1780s eventually reaching a depth of 202yd. The minerals worked included five seams of coal, two of ironstone and two of fireclay. These were all worked prior to 1870, from which time the mine concentrated on two additional seams of red tile clay, which were worked from an inset of about 127yd down the shafts.

The site now forms part of the Blists Hill Museum, the engine house has been rebuilt and a steam winding engine from nearby Milburgh Mine, similar to the original engine, installed. The headframe has also been reconstructed and the shaft re-opened for about 15yd depth. Winding from this depth is once again possible.

21 *(above)* Showing most of the miners employed and the two ponies Ben and Turpin about 1890. These ponies were raised and lowered through the shaft at this and the adjoining Shawfield Mine as required. Since 1870 the mine rarely employed more than twenty men at any one time. The early engines were of the Heslop type but from 1912 a horizontal single-cylinder engine was used. This remained at the mine until 1956. *(The late W. Harper)*

22 *(opposite, top left)* The mine cage at surface in 1935. The small tub contained a load of hard red clay which had to be weathered before it could be used. *(F. Turner)*

23 *(opposite, top right)* Water pit, Blists Hill, about 1940. The photograph shows how the mine was kept clear of water. A bowk or kibble was lowered into the sump, filled, raised to the surface and emptied into a wooden trough, which passed the water into the nearby canal. *(F. Turner)*

24 *(opposite, bottom)* Watering, weathering and turning the clay about 1935. As the clay was very hard when first brought to surface it had to be weathered for three months before it could be used. This consisted of stacking, watering (a spray can be seen to the right of the picture) and turning regularly. The local policeman and a couple of friends kept themselves fit by doing the regular turning under contract. *(F. Turner)*

25 The coke hearths, Blists Hill about 1890. The mine headgear, engine house and haystack boiler can be seen in the background. In the foreground the coal is being coked by stacking around small pillars, covering and partially firing. The coke was used in the adjoining blast furnaces. *(B. Waterson)*

The Tar Tunnel (SJ 694025)
This tunnel was constructed from the north bank of the River Severn at Coalport beneath Blists Hill towards Madeley in 1787. The object appears to have been to assist with the transportation of mineral from the Blists Hill and other mines of the Madeley Wood Company to the River Severn. The tunnel connects with the Blists Hill Shaft at about 60yd down and in view of the close proximity of the shaft to the canal, which runs on on the surface parallel with the tunnel, it seems quite likely that it was part of a surface canal, shaft and underground canal system which predated the use of canal inclined planes. Similar systems were used locally at Coalbrookdale and Lilleshall.

Soon after the tunnel was commenced a natural spring of bitumen was encountered and produced commercially. This bitumen, or 'tar' as it is called locally, can still be seen exuding from the walls of the tunnel which has now been re-opened for part of its length as a feature of the Blists Hill Museum. The tunnel was re-discovered and surveyed in 1966 by members of the Shropshire Mining Club and has been explored for a distance exceeding 700yd.

26 A view in the tunnel in 1970 before it was cleaned out and opened to the public. The black streaks on the tunnel sides were caused by the bitumen running down the walls.

27 A culverted drain runs the full length of the tunnel and where open can be seen to be carrying substantial quantities of 'tar'. The photograph shows a well where the bitumen was collected. *(Institute of Geological Sciences)*

28 **Madeley Wood Colliery,** actual location not known. This engine could have been at any of the mines in the Madeley Wood area. It is a Heslop beam winding and pumping engine and the photograph was taken in 1879. In 1878 an engine of this kind, and then believed to be the last, was presented to the patents Museum (now the Science Museum) London. The notice of this donation in *The Times* of 24 May 1879 led to a letter from W. R. Anstice of the Madeley Wood Company saying that they 'still have three of the same description at work, they had five and once had eight'. The last of these engines ceased work in 1890. *(Science Museum, London)*

29 and 30 *(opposite)* **Lane Pit, Madeley,** about 1890. Although the actual location of this mine is not known for certain, it was probably the 'old shaft' in the field called Old Lane Pit Leasow (SJ 680042) near Hill Top Farm on Madeley Hill.

The photographs show a very primitive steam winding engine (and possibly pumping) engine, almost certainly of Heslop design. Records show that a steam engine was erected at this mine between 1794 and 1797. Prior to this date horse gins were used as is shown by an inventory taken at the 'Lane Pit, Madeley at Christmas 1747'. This inventory included twelve bank wagons at 6s, two gins at £10, two gins at £5, two pumps at £5, a pit chain at £14, twelve baskets at 6s, four shovels at 9d, six mandrils at 1s, ten augers at 5s, three hammers at 3s, eight dressers at 4d, eleven axes at 1s 3d, two horses 'in the pit' at £4, 2 tons of hay at £3, 350yd rail 'in the pit', 200yd rail 'on the bank'. There was also a further 400yd of rail which presumably connected the mine with the 'store house' and 'pens' by the River Severn.

It is interesting to compare these pictures with Fig 28 for there are many similarities, and it is probably the same engine photographed at a later date. *(IGMT)*

Shawfield Mine (SJ 698037) The mine was probably sunk about 1840 and until final closure about 1916 remained in the ownership of the Madeley Wood Company. In later years it was worked alternately, month by month, with Blists Hill Mine. During this later period the mine produced mainly fireclay from the Productive Measures.

31 and 32 General view of pumping and winding engines in 1917. The steam winding and pumping engine was of the atmospheric type with a single open-ended vertical cylinder, 3ft diameter by 4ft stroke. Steam was supplied at 4lb per sq in from the egg-ended boiler. This can be seen to the right of the house, but before this boiler was installed the haystack boiler on the extreme right had been used. The 14ft long beam was of oak and drove the winding gear (winding chain was used) through a crank. *(The Engineer)*
(A full description of the engine is given in *The Engineer*, 3 Aug 1917).

33 **Shaft sinking at Shawfield Mine,** Madeley, 1930. It was common practice in the coalfield to cover surface streams with brick arching and then form spoil heaps above. At the Shawfield Mine in the late 1920s such a culvert collapsed, damming the Wash Brook and causing extensive flooding upstream. The landowners, and operators of the adjoining Blists Hill Mines, George Legge & Son, were sued for damages by the local council and went bankrupt in 1934. In the meantime an attempt was made at re-opening the culvert by sinking a 70ft deep shaft through the spoil using a handwinch (see photograph) and forming a bypass tunnel around the collapsed section. *(B. Waterson)*

34 **Transportation of coal by tub-boat from Kemberton Mine** (SJ 701049), about 1900. This section of canal, opened in 1792, was severed from the main coalfield tub-boat system in the 1850s and from this time connected only the mines of the Madeley Wood Company to their brickworks and blast furnaces and, via the Hay Inclined Plane, to the Coalport China Works and River Severn. The canal was extensively used for mineral transport until the turn of the century and finally closed in 1916. In the background can be seen the Madeley Court Company's blast furnaces and to the left the stack of one of that company's mines. W. Richards was the last tub-boat bargee. He is seen in the photograph holding the pole which prevents the boat from hitting the canal bank. *(The late W. Harper)*

35 **Part of the Madeley Court Colliery** (SJ 695048), about 1906. Extraction of coal, ironstone and fireclay commenced in the vicinity of the Elizabethan Madeley Court or Manor House about 1840 and in 1843 blast furnaces were constructed by James Foster. The furnaces continued to operate until 1902 and were supplied with mineral from shafts numbered from 1 to 17. All the mines were worked under the chartermaster system. Ironstone production ceased in 1902 but some coal was produced until about 1910. The shafts varied in depth from 166yd to 300yd. In 1882 the surface equipment consisted of one pumping engine, seven winding engines and one 'old water engine', all were steam powered and of the beam-type. In 1864 the beam of one of the engines broke and fell down the shaft completely blocking it. Fortunately the miners were able to escape through a second shaft, unlike the great Hartley Colliery disaster in Northumberland two years previously when an identical accident blocked the only shaft and 204 men were buried alive. *(The late W. Harper)*

THE LILLESHALL COMPANY

The Lilleshall Company was formed in 1802 to continue the work of the earlier partnership of the Marquis of Stafford and the Gilbert brothers, who prior to this time were content to act as ground landlords to the mineral producers. Like the Madeley Wood Company this company was multi-industry based, but on a larger scale, operating coal, ironstone, fireclay and limestone mines, blast furnaces, forges, brickworks and works for other clay products etc. The company is still operating (1975) a rolling mill, steel fabrication works and concrete brickmaking plant.

In 1871 the Lilleshall Company operated forty mines and produced 400,000 tons of coal, 105,000 tons of clay-band ironstone, 5,000 tons of fireclay and over 10,000 tons of limestone. The company had fifteen pumping engines and raised nearly 2½ million tons of water per month. Many of the mines used triple-linked winding chain with skips but a report of this time states that they were introducing cages in order to increase production. The company made nearly all its own engines and patented an early mine-ventilating fan. Both engines and fans were supplied to other coalfields. By 1891 the company was operating fifteen mines, in 1917 only five and at Nationalisation in 1947 only two, the Granville and Grange, both near Donnington.

Stafford Mine (SJ 715091). The mine's history is well documented, the first sod for the shafts was cut on 24 November 1862 after a ceremonial scattering of 'silver dewdrops'. Despite this it was not until 8 December 1866, and many tribulations later, that the intended depth of 255yd was reached. The mine expanded to become one of the largest in the coalfield, employing over 400 men by 1905, but, after ironstone working ceased, numbers fell until by 1924 only 270 remained. Coal production ceased in 1926 but pumping continued until 1937.

36 This photograph is taken from a geological section by T. Doody of about 1870 and shows the general surface arrangements when the mine was opened. The engines included a Bull-type engine with 45in diameter vertical cylinder, 7ft 6in stroke for pumping. The piston rod was directly coupled to the pump rods, but the engine was assisted by a large outside beam which had a balancing weight in an iron tank. The winders were of the rotative beam type, made by the Lilleshall Company. One survived until the mine closed, this had a 36in bore cylinder with 6ft stroke and a 15ft diameter winding drum and 18ft diameter flywheel. It operated on a pressure of 30 lb per sq in. In later years the other winder was changed to a horizontal-cylinder type with a crank on the rope-drum shaft.

STAFFORD PITS NEAR SHIFFNAL, SHROPSHIRE.

37 A close-up of the rotative beam-type winding engine taken about 1930. All the engines at the mine were scrapped at the end of World War II. *(G. Watkins)*

Granville Mine (SJ 725120). An early geological section shows the shafts to have been sunk to the lowest workable coal seam, the Lancashire Ladies Coal at 409yd, by 1860. In the 1950s one of the shafts was deepened to 444yd to become the deepest in the coalfield. According to the Lilleshall Company's former engineer, W. Atkinson, rotative beam-winding engines were used at both shafts. No 1 Pit had a single vertical cylinder 32in diameter and 5ft stroke connected to the rope drum by a crank. Two flat ropes were used for winding, each rope raising coal from a different depth. No 2 Pit had a pair of engines with two vertical cylinders each 16in diameter and two flat-rope drums. The mine also had a steam pumping engine of the Cornish type, 74in cylinder, 10ft stroke with steam pressure at 24lb per sq in and a 20ft diameter steam-powered Guibal fan which was in use before 1890 and replaced a furnace in the shaft bottom.

This mine was the last in Shropshire to work under the chartermaster system (a Mr Cooper, who retired in 1913). The mine normally employed about 300 men but reached a peak of nearly 900 men in 1967 following the transfer of men from the Kemberton Mine. After Nationalisation a full modernisation programme was carried out, all the steam engines were scrapped and electric winders installed in 1949-50. These were the first such installations carried out by the National Coal Board in the country. Later a further programme was undertaken, costing over a million pounds, which included the widening (to 16ft diameter) and deepening of one of the shafts. Twelve-ton electric battery locomotives were introduced underground to haul 2½ton capacity minecars on the 2ft 3½in gauge track. The present coal faces are about 2 miles from the shafts and the coal is won by shearer cutting and loading machines. The mine has many geological problems and for some years has been under the threat of closure, although at present (1975) the future looks more promising.

38 The Cornish pumping engine and vertical winder in 1864. From a geological section by T. Doody.

39 A similar view to Fig 38, but a second vertical engine has been added. From a painting by R. Williams dated 1879 in the Surveyor's Office, Granville Colliery. *(NCB)*

40 A further view of Granville Mine in 1944. It can be seen that few changes occurred after 1879. *(IGMT, Lilleshall Collection)*

41 General view during reconstruction about 1949. Note the new steel headframe constructed around the old one. *(NCB)*

42 Granville Mine in 1964.

43 The underground haulage road with main and tail rope haulage system in 1938. *(IGMT, Lilleshall Collection)*

44 *(opposite, top right)* The winding arrangements at Granville Mine in 1945. A bi-cylindrical-conical drum is shown, a system which gives a torque advantage on commencing to wind. The engine and winding gear was constructed by the Lilleshall Company. *(IGMT, Lilleshall Collection)*

45 *(Opposite, top right)* No 2 winding engine control gear, 1950. *(NCB)*

46 *(opposite, bottom)* No 2 Pit winding 'spool' or drum for flat rope, 1950 *(NCB)*

Grange Mine (SJ 720114). The Lilleshall Company commenced to sink the shafts of this mine in 1864 and eventually reached a depth of 320yd. The winding engine had a single horizontal cylinder 45in diameter connected to a flat-rope drum. This engine replaced an earlier engine in 1871. The pump was similar to that at Stafford Mine. It was of the Bull-engine type, 42in cylinder, stroke 9ft, 40lb per sq in steam pressure and was non-condensing. The engine was built at the company's works. In 1891 the mine also had a steam-powered 20ft diameter Guibal fan. The engines were scrapped in 1951 when electrical equipment was installed and the mine was connected underground to the Granville Mine.

In 1947 Grange Mine employed 150 men but ceased production as a separate unit in April 1952. It is now used as upcast ventilation shafts for Granville Mine and has a methane drainage plant attached. This plant is at present discharging some 600cu ft per minute of almost pure methane to the atmosphere. The gas is obtained from the coal seams at Granville Mine and was formerly sold to the local Gas Board.

47 The twin wooden headframes in 1950. *(T. H. Stonehouse)*

48 Photograph taken during the reconstruction of the headframe in 1951, the steel tandem headframe encompassing the old timber frames. *(T. H. Stonehouse)*

49 The tandem headframe at Grange Mine when completed, 1964. The shafts are enclosed because they serve as upcast shafts for the Granville Mine. To the right can be seen the chimney for exhausting to atmosphere the methane drained from the mineworkings.

50 An unusual view of a flat winding-rope in use, 1951. The cage and wooden tubs can also be seen. *(T. H. Stonehouse)*

51 The 'spool' winding drum for the flat rope and horizontal steam winding engine, 1951. *(IGMT)*

52 and 53 **Waxhill Barracks Mine** (SJ 718129). One of the earlier Lilleshall Company mines dating from the 1820s and with a depth exceeding 300yd. The mine worked coal and ironstone and ceased production about 1900, although pumping continued until 1930. The pumping engine worked on the Cornish principle, it had a 61in diameter vertical cylinder and was double-acting. The engine was scrapped about 1940.

The photograph was taken shortly before the mine closed and although of very poor quality it is the only known illustration of Shropshire pit-women in their characteristic dress. The drawing by M. Newton shows details of the costume. Several thousand such women were employed at one time on the coal screening plants and on the spoilheaps picking and carrying ironstone nodules. *(W. Howells)*

54 **Freehold Mine** (SJ 717134). Another early Lilleshall Company mine probably sunk in the first quarter of the nineteenth century. The final depth was 245yd. The mine ceased production in 1928 owing to very heavy pumping costs. This mine worked coal from some of the steepest seams in the coalfield, at times lying at more than 45 degrees to the horizontal. The photograph shows the mine shortly before it closed. It is likely that the tank or reservoir seen towards the centre was originally a canal tub-boat. This was a common use for such boats and one has only recently ceased to serve this function on a nearby farm. This boat is now afloat again on the Blists Hill Museum canal. *(R. Tranter)*

Woodhouse Mine (SJ 713103). In effect two mines each having a pair of shafts sunk by the Lilleshall Company, probably in the second quarter of the nineteenth century. The mines worked coal and ironstone from shafts up to 311yd deep. During the 1920s this became a very large mining complex for this coalfield, employing over 740 persons in 1922. For ventilation the mine originally had a Lloyd fan which was patented at the company's works. This was 16ft diameter and 5ft wide, but by 1891 it had been replaced by a 28ft diameter Guibal fan running at 70 rpm. No 1 Pit had a winding engine with vertical cylinder of 32in diameter and 5ft stroke, connected by a crank to a rope-drum shaft and two flat-ropes. The drum was directly coupled to the crankshaft, and the pistons went straight through the cylinders. No 2 Pit winding engine had a pair of horizontal cylinders with cranks on the drum shaft. No 1 Pit ceased production in 1931, No 2 Pit closed in 1940 but some pumping was done until the 1960s using electric pumps. All steam engines were scrapped during the 1940s.

55 From an unsigned painting in the Surveyor's Office, Granville Colliery. This is believed to be of the Woodhouse Mine about 1880 but it could be some other Lilleshall Company mine. It is interesting to note that the engine is at normal ground level while the shaft top and headframe have been built up as spoil has been brought from the mine. Several local mines show this unusual feature. *(NCB)*

56 This photograph, believed to be of No 2 Pit, shows the large engine house and steel headgear used in later years. The photograph was taken in 1960. *(P. J. Donnelly)*

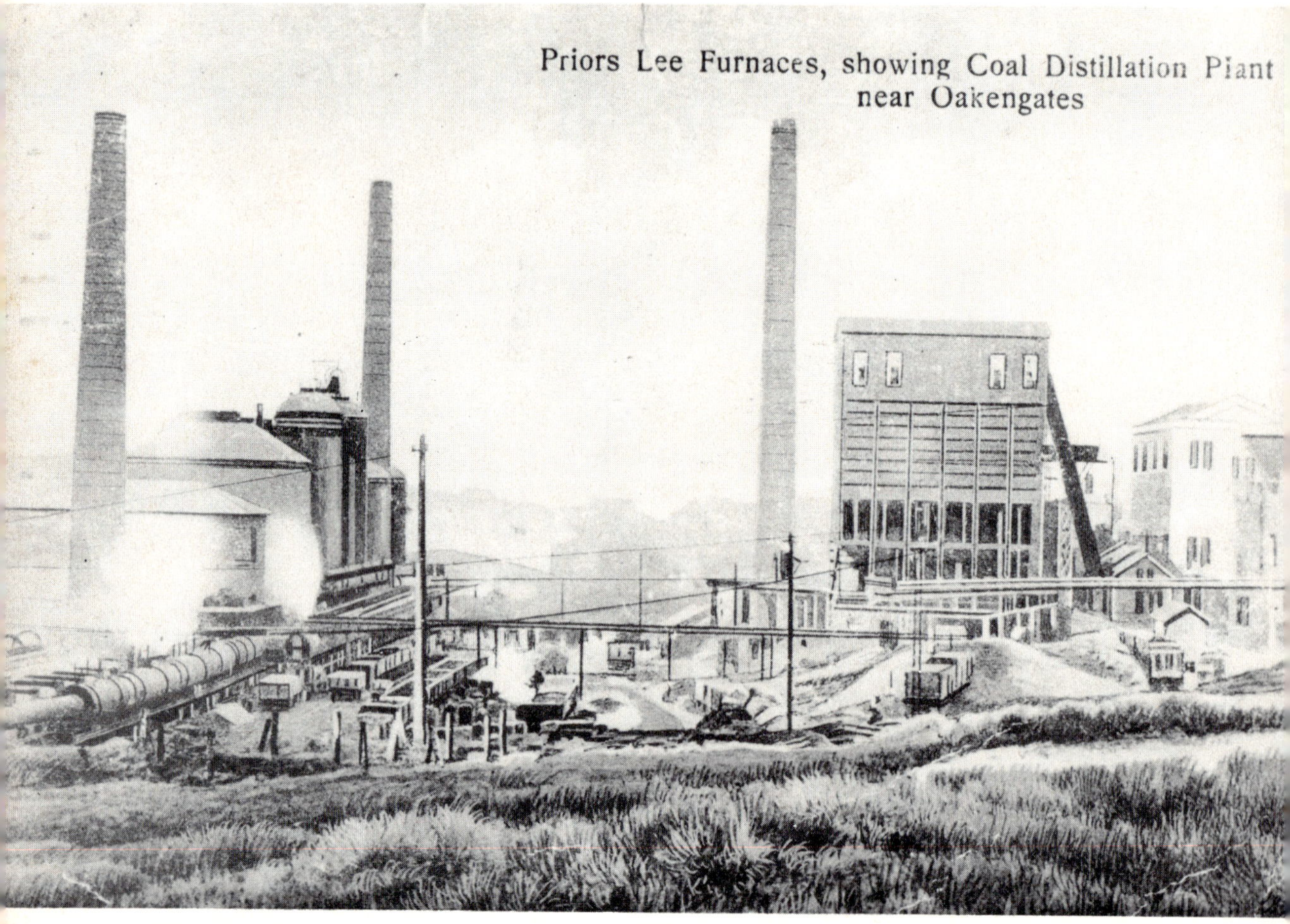

57 **Coal distillation plant at Priorslee** (SJ 701099) probably about 1890. There has been a long history in the coalfield of the extraction of useful products from coal and shale. Pitch was being extracted as early as the 1690s in Broseley and, a century later, most of the local ironmasters had plants for the production of coke and tar. Earl Dundonald's 'British Tar Company' operated plants at the Calcutts and Benthall until the early years of the nineteenth century, at which time there were plants at nearly all the principal furnaces.

Little is known of the plant at the Lilleshall Company's Priorslee Furnaces but, taking the photograph as an indication, it must have been a considerable one. The plant was situated between the furnaces (blown out in 1959) on the left of the picture and the main Holyhead Road (A 5). *(D. Jobber)*

58 **The Nabb Pit** in 1912 (SJ 699114). Miners on strike during the 1912 dispute. Whenever the miners were not working at the larger mines they opened up their own 'jackey pits'. The photograph shows such a mine. An upturned bicycle, an old bath and a riddle was all the equipment required. The mine was at The Nabb, St Georges. It was 7yd deep and the leader was Harry Edwards (on the left, with a pick over his shoulder). The Nabb is an area of very shallow coal seams and there is some evidence to indicate that the Romans worked the Fungus Coal Seam here when constructing their Watling Street.

59 and 60 *(overleaf)* **Ivy Colliery,** Clares Lane, Ketley, in 1934 (SJ 686095). A small mine owned by Edward Harris (in dark suit). The pit was 15yd deep and produced 36 tons coal per week. The owners of this mine were among the first to use Bedford trucks for coal distribution and the photographs were taken by that company to advertise their trucks. Worked from 1926 to 1939 the mine employed no more than thirteen men at any one time. As can be seen the hand-winch drum was so large that it had to be worked through gears. *(A. Harris, Lawley Bank)*

AMW 583

UX 9610

61 **Harris's Old Park Pits** (SJ 685088). A general view taken just after Nationalisation in 1947. The mine employed about twelve men including Albert Harris, former owner, his three sons George, Len and Arthur (Arthur built the headframe). Also employed were three brothers named Lord. The mine was closed about 1950. The steel headframe was moved to Farm Mine, near Heath Hill, and more recently to the Blists Hill Museum. The winding engine was electrically operated. *(G. Harris, Wellington)*

62 **Brandlee Mine** in 1962 (SJ 678076). The Brandlee is a very old shallow mining area with numerous shafts in a complex which extends over Horsehay Common to Spring Village. Mines on this site, named 'The Brandlee', were at work throughout the last century but the mine shown was owned and operated by the Tarr family. One shaft was sunk during World War I to the Clod and Randle Seams at about 33yd. The other shaft was sunk to the Big Flint Seam at 8yd and later to the New Mine Seam at 13yd. Other shafts were sunk as the need arose. The corrugated-sheet covered house contained a horizontal steam engine and vertical boiler which was used until the mine closed in 1956. This was the last steam-operated winder to work in Shropshire and it remained in position until it was scrapped in 1962.

Shortwoods Mine in 1967 (SJ 658096). Situated on a site with a very long history of mining, the adit mine shown closed as late as February 1970, the last of the thousands of private mines to have operated in this coalfield. It was partly owned by Mr. H. Tarr who was also the manager. The adit shown was No 10 which was constructed in 1953. Most of the other adits at this mine were still open and two were used for additional access and ventilation. The adit was driven from the outcrop of the Two Feet and Best Coal Seams, which lie together at this point, and have a combined thickness of 3ft.

The pillar and stall method of working was used, the depth being only 23yd at maximum. A 'Siskol' arc wall shearer undercut the Best Coal Seam in the 12ft wide stall to a depth of 4ft 6in. The coal was then blasted and hand loaded into 10cwt capacity wooden wagons, which were then hauled by small electric winches to the surface. At the surface the wagons were hauled up a ramp and tipped manually onto a screen. The larger coal passing this was further sorted using 'franks', five or eight pronged forks as seen in the photograph. The mine employed about twenty-five men and produced 200 tons of coal per week.

63 *(opposite, top left)* The adit entrance to the mine.

64 *(opposite, top right)* The ramp from the adit to the top of the screening plant. At the top the tubs were tipped on their side by hand.

65 *(opposite, bottom left)* Hand screening of the coal using 'franks' or forks.

66 *(opposite, bottom right)* The bins and loading bays.

67 **View of Jackfield Clay Industry,** about 1898 (approx SJ 682029). The principal works shown is believed to be that of W. & P. Jones, and known as the Hollygrove Brick and Tile Works. The company operated its own mines for red clay in the vicinity of the old church at Broseley Wood. A horse gin at one of these mines can be seen on the horizon. The shafts were at The Hollygrove and The Rock and were about 33yd deep. Clay was brought to the surface at the works, further down the hillside, by means of an adit. *(R. Miles)*

Guest's Deep Mine, Broseley (SJ 683016). A very old mine and probably the mine described as 'a Deep Pit on Madl. Crompton's land' in a deed of 1726. The shafts were about 150yd deep and while in later years the mine was mainly a red clay mine it had also produced some white clays and a little coal. In later years a 'fat grey glacial clay' was also produced from a quarry near the shaft top. This was added to the red clay for tile making in the proportion two of red to one of grey. The mine rarely employed more than eight men who produced about 24 tons of clay per week. The clay was hand got and loaded into 10cwt trucks which were then trammed to the pit bottom. It took five minutes for the engine to raise each tub. The mine closed in 1941, up to which time it continued to be furnace ventilated.

The engine was of great antiquity, the design and some parts were believed to be 130 years old. It consisted of a single vertical cylinder of 15in bore, 38in stroke and was double-acting. The piston was connected to a 10ft 6in long cast-iron beam pivoted at the centre. The crank transmitted power to the two cable drums, of different diameters to wind from different depths in the shaft, which were geared in as required. The engine worked on a pressure of 5lb per sq in, steam being produced originally in a haystack boiler but later an egg-ended boiler. Attempts were made to preserve the engine, but since 'it represented no particular period of development'(?) these failed and it was scrapped in 1945.

68 General view of the surface about 1930. It is interesting to see that low trucks or 'dans' having plate wheels, the flanges being on the rails themselves, were used to take the clay to the heaps for weathering. The photograph is taken from trade literature of that period. *(IGMT)*

69 *(above)* Close-up of a mine 'dan' with plate wheels, about 1936. *(P. Kendrick)*

70 *(left)* The mine cage suspended on four chains. The shaft was fenced for safety after the mine closed. *(F. R. Gameson)*

71 and 72 The winding drums, flywheel and wooden engine house of Guest's Deep Mine in the 1940s. *(A. G. Sidebotham and F. R. Gameson)*

Rock Fireclay Mine (SJ 680092). Originally opened for coal in 1882 and over the years since then more than a dozen shafts have been sunk at the mine. James Jones & Sons took over the mine in 1928 to work coal, fireclay and iron-pyrites but after 1956 only fireclay was produced. At the time that the mine finally closed in December 1964 there was an electric winder and a handwinch at the two fireclay shafts respectively, a motorised hand-winch on one of the coal shafts and two hand winches at the other.

In later years the Clunch fireclay seam was worked from a 30 yd-deep shaft using the pillar and stall system, the pillars being about 20ft square and the stalls or headings 6ft wide. The pillars were then robbed by driving further headings through them, known as 'splitting the pillars'. Hand boring machines were used for drilling shotholes and, after blasting, the clay was hand loaded and trammed to the pit bottom. Candles were in use for illumination and as a protection against running into accumulations of blackdamp. The mine produced 150 tons per week and employed ten men.

After the mine closed it was salvaged by the Shropshire Mining Club. In order to get the ventilation circulating at the coal shafts a fire bucket had to be suspended down the shaft using the cast-iron handwinch. A Siskol coal cutter, purchased in 1937, was removed from the mine together with much other equipment dating from this time. This material is now at the Blists Hill Museum.

73 The main fireclay shaft headframe and surface buildings in 1964.

74 The main coal shaft motorised-handwinch and suspended box in 1965. The box was in fact the top of a mine tub. When carrying mineral all three chains, two at one end, one at the other, were attached so that the box was level. When carrying persons the box was suspended from the two chains only so that it was at an angle as in the photograph. The rider then stood on the end of the box and held the winding rope. In the foreground is the 'tacking plate' a platform on wheels, which was pushed over the shaft and loaded boxes were lowered onto it.

75 The firebucket and handwinch at the No 2 coal mine in 1965. This is the last known occasion that a firebucket was lowered in a shaft in Shropshire, and almost certainly the last in Britain.

Lincoln Hill Limestone Mine (SJ 669038). The Silurian Limestone has been worked since at least the early seventeenth century and by 1801 there were 'several prodigious caverns of limestone supported by stupendous pillars'. The mine was worked by adits from the side of the Severn Gorge and from several 30-45yd deep shafts from the top of the hill. During the late ninteenth century the vast pillar and stall workings were opened to the public and were included on the list of attractions of the old Severn Valley Railway. The mining of limestone was continuous over at least a century and finally ceased in 1907. During later years only three men were employed producing about 600 tons a year. The beds were 30-40ft thick and they dipped at an angle of 35 degrees. This inclination appears to add height and vastness to the workings which are still accessible to potholers. Donkeys were used for haulage underground, but little evidence has been found in the mines of the use of anything other than the simplest tools, shovels and wedges.

77 *(opposite, top)* Adit entrance at Lincoln Hill on a hillside ledge above a number of ruined limekilns. The adit can be entered for 35yd where it enters a limestone shaft at a depth of about 50ft. This may in fact be a very large kiln shaft and firing tunnel but this possibility is still being investigated.

78 *(opposite, bottom)* Limestone shaft headframe and winding chain near the top of Lincoln Hill in 1907 (SJ 670038). The shaft is 61yd deep and was last used early this century. It is believed that the shaft was sunk about 1800 and, according to J. Prestwich, a local geologist, passed through the lower seams of the Coal Measures. *(D. Oakley)*

76 Photograph taken in Lincoln Hill Mine in 1975. A member of the Shropshire Mining Club is standing on a heap of debris looking along the strike of the limestone bed towards a solid pillar of rock. These pillars are 20-30ft high and about 20ft square. Fourteen such pillars support an area of working measuring about 110yd x 50yd. The mine is now only accessible to potholers who have to climb down a 36yd shaft. *(D. Stevenson)*

OLD 'GIN' PIT!
NEAR LIGHTMOOR

79 *(opposite, top)* **Horse Gin at Deep Pit, Broseley,** about 1940. This photograph is the most recent available of a horse gin in the coalfield. The gin is interesting because it was constructed very simply, of unbarked timbers, and it had only one rope which was wound on and off the drum as required. It is known that at least one horsegin was in use in the coalfield near Madeley as recently as 1948. *(P. Kendrick)*

80 *(opposite, bottom)* **Horse Gin at Lightmoor,** about 1937. This drawing by J. Willocks is from a photograph which appeared in *Shropshire, the geography of the county* by W. W. Watts, published in 1939. This was a fairly common type of horse gin with two ropes, one over each pulley shown on the left of the picture. The two ropes each had a weight, normally a box, one would go down the shaft while the other went up. The auxiliary pulleys prevented the rope dropping off the main drum when one of these weights was removed. Note the unusually narrow rope-drum.

81 **John Randall FGS** (1810-1910). Born in Broseley, Randall lived to a great age and survived until the early years of this century, consequently seeing many changes. Although he had no formal geological or mining training he taught himself the subjects and was, in later years, often called in as a consultant in mining geology. In addition he was an accomplished writer and produced several books including works on the clay industry and district histories, all containing much geological information. He was also a skilled bird painter in ceramics having worked for the celebrated Coalport China Company and in the Rockingham Works. 'To his geological researches was due the extension of the mining industry in this neighbourhood.' (Epitaph, Madeley Church)

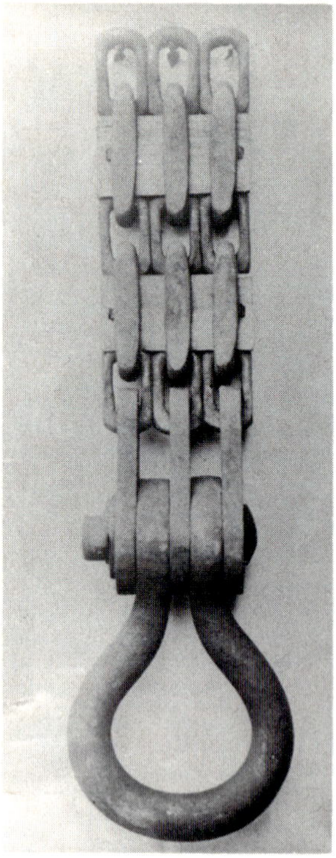

82 **Shropshire Winding Chain.** Chains seem to have been used for winding since at least 1804 and by 1807 the Botfield concerns had 2,000ft in 'daily motion'. By 1814 G. Gilpin was claiming that the Lilleshall Company alone used 8,000yd of his pit chain. Gilpin had received a Society of Arts prize for his work on pit chain in 1805. By 1810 the triple-linked chain had been developed using wooden pegs for holding the links together. This improved safety considerably. This type of chain won the local firm of Edges several awards, including one at the 1851 Exhibition. Winding chains continued to be used locally until after World War I, but from 1870 they were gradually superceded by flat-wire ropes. Considerable lengths of this chain can be found in the coalfield being used as fencing. The firm of Edges, later part of British Ropes Ltd, closed in 1971. The photograph shown was taken by George Maw about 1861. *(IGMT)*

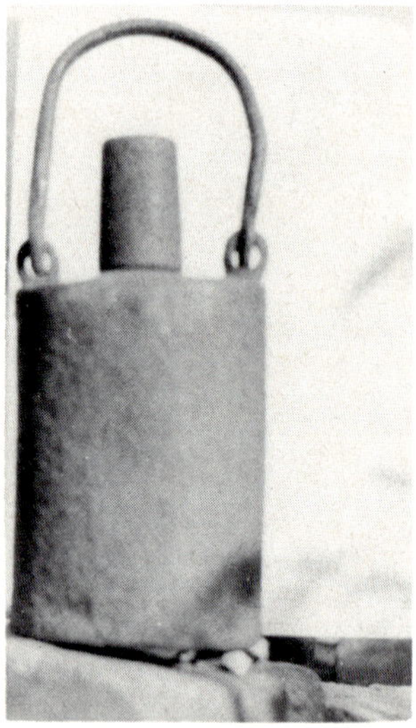

83 **Powder Can and Measure** found in old mine workings at Broseley. Obviously made by the local blacksmith. Used for carrying blackpowder into the mine for blasting purposes.

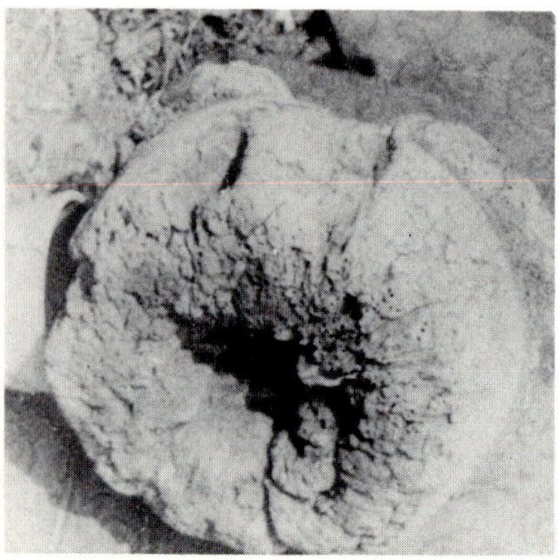

84 **Wooden Flanged Wagon Wheel.** This wheel was found in a Broseley mine about 1930 and is something of a mystery. Wooden railways had been used in Shropshire since the early years of the 1600s. These early rails were of the flanged type and the wheels which ran on them were flat discs of wood. During the 1720s the Coalbrookdale Company began to cast iron-plate wheels and in 1767 began to use cast-iron rails also and later flanged rails were introduced. Similar metal flanged rails and plate wheels were in use in Shropshire mines until the 1950s (see Fig 69). The transition from wood to metal, and of the flange from rail to wheel was obviously a slow process, but it is most odd that early wooden wheels such as the one in the photograph should have been used having the later flanged shape.

85 **The Lloyds Pumping Engine** about 1910 (SJ 689030). A 'central' pumping engine used for draining the mines of the Madeley Wood Company. The pits were probably sunk in the late eighteenth century and a steam pumping-engine is indicated here on a plan of 1840. A further plan of 1882 shows that the engine position had been changed. Recent research indicates that the early engine was of the Newcomen-type, probably replaced by a Watt engine in the middle of the last century. The engine ceased work in 1912 although the pump spear still stands about 14ft high out of the shaft fill. Some site investigation work has recently been carried out by the Ironbridge Gorge Museum and it appears that there are substantial remains of the pit work and engine foundations just below the surface. The shaft would have been about 80yd deep. No details are available of the engine but it was obviously of the 'Cornish' type. *(R.Miles)*

86 **Wombridge Pumping Engine House** (SJ 692121). Erected in 1858 this housed a 250hp Cornish beam engine with a 60in diameter cylinder, which could raise 3,388 gallons of water per minute with three strokes, although it was capable of up to 11 strokes per minute. The cylinder stroke was 10ft and pump stroke 9ft. Shortly after 1884 the engine was stopped because of a dispute between local mine-operators and the mines flooded. It is believed that the engine was then bought jointly by Haybridge Iron Co and Lilleshall Co and installed at the Pudley Hill Mine, close to the present Telford Centre. The mine was 175yd deep to the lowest coal seam but old strata sections showed that they reached, and almost certainly worked limestone at 197yd. Local tradition states that miners at one of the pits, the Cats Head Mine, had to run from the incoming water when the pump stopped. This engine house is well remembered by travellers through the area as it dominated the northern skyline until 1964. It was visible from both the main Chester to Didcot railway line and the Watling Street (A5) in the vicinity of Oakengates.

87 *(above, left)* **Lilleshall Company Barrack Houses.** Several long rows of single-storied, double-roomed barrack houses were built in the mining areas to provide accommodation for mineworkers. They were built during the late eighteenth century and continued to be used until the 1950s. The photograph shows Donnington Barracks and was taken about 1933. Barrack houses were also located at Waxhill, Lilleshall and several other locations. At Waxhill Barracks the 1851 Census shows that twenty of the twenty-seven single-storey houses had five or more inhabitants and nine had seven or more.

88 *(above, right)* **Priorslee Village** (SJ 710095). A small village constructed about 1839 for mineworkers at the Lilleshall Company's Lawn and Rookery Mines. This village is now a 'Conservation Area' and the cottages are to be modernised. *(The late W. H. Williams).*

89 **Limestone Miners' Cottage, Church Aston** (SJ 739177). This seventeenth-century thatched 'one up-one down' cottage is now maintained by The Shropshire Mining Club, a group of mining and caving enthusiasts who use is as a base from which to record and explore the mines of Shropshire and the surrounding counties. Recent repair has located several mining relics in the actual structure, including the use of flanged rail as door lintels. The artifacts shown around the building have all been salvaged from local mines. In the foreground lies the remains of a horse gin centre-post.

90 **Tramway tunnel with capstone dated 1800** (SJ 668036). Many tunnels were constructed in the coalfield during the eighteenth and nineteenth centuries for a wide variety of uses. These include underground tramways and canals, railway tunnels and others for which uses have not yet been discerned. One of the greatest tunnel builders was William Reynolds (1758-1803) who was said to be 'tunnel mad'. At least two of these are easily accessible today — this tunnel and the Tar Tunnel. The 1800 Tunnel is accessible for 340ft and is brick-arched throughout.

91 **Canal Tunnel, Wombridge** (SJ 691115). An inexplicable tunnel near the end of Reynold's canal of 1788 and in level ground close to the church at Wombridge. This tunnel was located in 1965 during building works and has now been backfilled. *(The late W. H. Williams)*

92 **'Result of the Coal Strike',** believed to be about 1912. This and other strikes brought much hardship to the miners and their families. This photograph shows family groups setting out from Cape Street in Broseley to seek outcrop coal in the surrounding countryside. *(D. Oakley)*

93 **The Coal Strike,** about 1912. A group taken at one of the outcrop workings at Foxholes, Benthall. The portly lady in the centre is obviously playing a leading role, as she holds both a sack and a riddle. The boy sits on a sledge but the tool he is holding is not easily identifiable. *(D. Oakley)*

94 **'Explosion and Fire at Shifnal',** 1821. A lithograph by Francis Nicholson (1753-1844). This obviously represents a mining scene in the Coalbrookdale Coalfield and is believed to show an explosion, probably of a steam engine or boiler. Accidents, like these and explosions of gas underground among others, were commonplace in this as in other coalfields. It would appear that the area was in some ways fortunate however, for no record has yet been found of any single incident involving the loss of more than twelve lives. *(H. M. Parker/IGMT)*

95 **'A view of Lincoln Hill with the Ironbridge in the distance taken from the side of the River Severn',** 1788, by James Fittler (1758-1835) after George Robertson (1748-88), published by I. and J. Boydell, London. This engraving appears to show the loading of coal lumps and slack into Severn trows for transportation down the river, possibly to Worcester and Gloucester. The River Severn was the main means of communication between the coalfield and the outside world up to the early years of the nineteenth century. *(Shropshire Star/Morley Tonkin Collection)*

Other Shropshire Coalfields

THE FOREST OF WYRE COALFIELD

Though of relatively little economic or historic importance this coalfield is the largest in Shropshire being about 50 square miles in area. It extends down the Severn valley from Bridgnorth to the Abberley Hills. Coal seams have been worked in both the Middle or Productive Coal Measures and the Upper Coal Measures. The coal seams in the Upper Coal Measures are sulphur-rich and generally of poor quality.

The lower Middle Coal Measure group of seams crop out to the west between Billingsley and Kinlet, but die out south of Highley. The principal seams are named from their general thickness: Five-foot (or Brooch) Coal, Halfyard Coal, Four-foot Coal and Two-foot Coal. The Upper Coal Measure group of seams have been worked only on a small scale on the Shropshire-Worcestershire border, but especially around Bayton and Mamble.

Highley Mine (SO 745830). The Highley Mining Company was formed in 1870 and began coal production in 1874. The Brooch Seam at about 300yd depth was worked throughout the life of the mine. In 1910 the company took over the 'Garden Village' which had been developed for miners from the once prosperous Billingsley Mine. In 1917 the Highley Mine was mechanised with the introduction of two Hopkinson Chain Coal Cutters. From the last quarter of the nineteenth century to the 1930s the mine generally employed 400 to 500 men. After the opening of the nearby Alveley Mine production at Highley ceased although the shafts were retained for ventilation and emergency egress.

96 *(opposite, top)* Drilling shotholes in the coal seam ready for blasting in 1919. The machine is a hand-operated screw drill and the operators are J. Jones (left) and W. Hall. Notice the candle in the 'gob' of clay; candles were in use underground until the 1920s. The screw machine replaced the old method of 'punching' shotholes with a handpunch. *(The late F. W. Beddoe)*

97 *(opposite, bottom)* General view of the mine in 1938 with plenty of evidence of steam engines. The main winding house was the large central building in the background. *(T. H. Stonehouse)*

98 *(opposite, top)* The twin steel headframes in 1938 with ropes worked from a common winder. *(T. H. Stonehouse)*

99 *(opposite, bottom)* A close-up of the 'outside' upcast shaft airlocks, 1938. This box-like arrangement ensured that the air was drawn through the mine by the fan and did not short circuit. *(T. H. Stonehouse)*

100 *(below)* Loading large lumps of coal into rail wagons by hand for the 'top-price' market, 1938. *(T. H. Stonehouse)*

Alveley Mine (SO 752842). In 1935-7 the Highley Mining Company sank a large shaft, about 377yd deep, on the opposite bank of the River Severn to their Highley Mine. This was to become the company's main production shaft. The mine was nationalised in 1947 and afterwards equipped with the latest machinery. In later years however the mine suffered geological difficulties, especially with regard to roof control. This prevented further mechanisation and the mine was closed as uneconomic in January 1969. During the mid-1950s the mine employed over 1,250 men, but this had fallen to 700 shortly before coal production ceased. The mine produced about 6,000 tons of coal per week.

101 *(above, left)* General view of the mine during sinking, 1937. This was the only mine to have a concrete headframe in Shropshire. *(T. H. Stonehouse)*

102 *(above, right)* A close-up of the shaft during sinking. Only the surface arrangements were square, the shaft itself was round. *(T. H. Stonehouse)*

103 *(below)* General view of the mine prior to closure 1967. The stores compound is in the foreground. *(N.C.B.)*

104 Aerial ropeway for carrying coal from Alveley shaft to the coal preparation plant situated on the opposite bank of the River Severn. Looking across the river towards Alveley, 1967. *(NCB)*

105 A coal face underground at Alveley, 1967. After power-loading using a shearer machine, the coal was transported on the conveyor in the foreground. The miners are advancing the roof supports towards the virgin seam on the left of the picture. *(NCB)*

106 Once off the face the coal was taken to the pit-bottom in minecars. The minecars in the picture were waiting to be loaded mechanically onto the shaft cage, 1967. *(NCB)*

107 The minecars were hauled by electric battery locomotives along the main haulage road as shown, 1967. *(NCB)*

108 A train load of miners waiting to be transported in carriages to points nearer their working places, 1967. *(NCB)*

109 **Bayton Mine** in the 1930s, (numerous locations about SO 685712). This mine, although just outside the county boundary had been included for completeness. Many mines, both shafts and adits, have been worked under this name in this remote border country and its miners are noted for their tough individuality. The largest mine in later years was Bayton No 6 which employed over seventy men in 1948, but was closed as 'uneconomic' by the National Coal Board in 1950. The No 6 Mine had been opened about 1938 to replace No 3, a mine worked on an ancient site from 1924. This earlier mine is said to have been started by a Mr W. L. Moody who paid his assistant in the early days not in cash, but in eggs and the occasional pig. The mines worked seams having the following names: Hard Mine, Five-feet, Main or Bank Farm Seams.

Hunt House Mine (SO 698701). Just over a mile from the county boundary, the Hunt House Mine was a modern private colliery. The first shaft was sunk by four brothers, G.R., E.T., P.H. and H.W. Mole about 1954. In 1960 the mine, operated by the aptly-named Mole Mining Ltd, employed thirty men underground and fifteen on surface and produced 30 to 50 tons per day. An electric winder was used to raise coal from the Five-feet Seam through the 61yd deep shaft.

When the company closed the mine in 1972 an attempt was made to work it as a miners' co-operative but the mine closed completely in early 1973. In October the mine was sold for £38,000 but, following the fuel crisis, has since been offered for sale again at £100,000.

110 *(right)* Coalface in the split seam, about 5ft high, 1970. Using a compressed-air pick to break down the coal. *(Hereford & Worcester County Museum)*

111 *(below)* The surface plant Hunt House Mine in 1970. *(Hereford & Worcester County Museum)*

112 The pit bottom looking inbye. *(Hereford & Worcester County Museum)*

113 The main return airway. *(Hereford & Worcester County Museum)*

THE CLEE HILL COALFIELDS

Two small isolated coalfields centred on the Brown Clee and the Titterstone Clee Hill respectively. The coal seams, lying immediately beneath the basalt (or 'dhustone') cap of each of the hills, have been worked from early times and the remains are the nearest Shropshire has to conventional 'bell-pit' workings. Some parts of these coalfields are over 1,700ft OD and for this reason they have been described as 'the highest in the Kingdom'.

The Brown Clee Coalfield is about two square miles in area and contains three principal coal seams, the Jewstone Black, the Threequarter and the Batty. All extraction appears to have ceased by 1850.

The Titterstone Clee Coalfield is about four square miles in area and is in turn made up of a number of smaller coalfields centred on the localities of Knowbury, Catherton, Cornbrook and Clee Hill Village. Five seams have been worked and these are in general thicker than those of Brown Clee. Some ironstone has also been worked and Carboniferous Limestone has been mined and quarried south of Clee Hill Village. The last coal mine in the coalfield closed in 1927. Limestone was last worked underground in the vicinity in 1914.

A characteristic feature of the miners of the South Shropshire coalfields was their part-time nature. Many, even in recent years, worked in the mines and at the same time ran smallholdings 'on the hills'. It was not uncommon for the smaller mines to have to close down at certain periods of the year such as lambing and harvest.

114 **Bell pits at Lubberland.** Catherton Common in 1955 (about SO 625779). *(Cambridge University Collection: copyright reserved)*

THE SHREWSBURY COALFIELD

This coalfield really comprises three separate coalfields, each much worked by small pits in the past. The Dryton Coalfield near Eaton Constantine is small and isolated and has not been worked for well over a century. The Hanwood Coalfield averages about three miles in width and extends for eighteen miles between Haughmond Hill and the Breidden Hills. The third coalfield at Leebotwood is about 30 square miles in area and is almost square in shape.

The coal seams are all in the Upper Coal Measures which, in this area, rest uncomfortably upon Pre-Carboniferous rocks. In the two more important coalfields the workable seams, in descending order, were the Half-yard Coal (about 1ft 6in thick), the Yard Coal (2ft 9in) and the Thin, Best or Deep Coal (1ft 6in). Associated with these seams there was a bed of limestone about 5ft thick, which, prior to 1870, was also mined and used for agriculture.

115 **Pontesford Mine Pumping Engine House** (SJ 410065). The collieries of this area were used to supply the fuel for the steam engines and smelters associated with the adjacent metal-mining area. As early as 1775 a 'fire engine' (or Newcomen-type of steam engine) is recorded in the vicinity and there is good documentary evidence of several other steam engines used at Pontesford. However it is not known which engine was installed in the house shown in the photograph. It is known however that this house has been a dwelling since before 1840 as the present occupier's grandfather was born in the house in 1843. The building is divided into three storeys and the engine pit is used as a cellar. The pumping shaft is about 6ft from the lever wall, but is now filled up. The boiler house has been incorporated into the local blacksmith's shop, which is built onto the west wall. Nearby is a further smaller engine house also converted into a three-storey dwelling.

Hanwood Mine (SJ 436093). Only one workable seam, believed to be the Half Yard Coal, was present at this mine. The seam was less than 2ft thick so that in haulage tunnels the roof had to be 'ripped' to give headroom, the material being put into the goaf. Sledges, pulled by a chain between the miners' legs, were used on the coal face until the late 1930s.

The single shaft, 140yd deep, was connected by workings to the disused Cruckmeole Mine nearby, whose shafts were also used for pumping and ventilation. Hanwood shaft was sunk in the 1870s and was purchased in 1920 by Mr A. N. Fielden, the mine was then worked as the Hanwood and Moat Hall Collieries Ltd until it closed in 1941. Between 1921 and 1941 numbers employed fell from over 300 to less than 100, but for most of this time production averaged 25,000 tons per year. In 1921 an electric generator was installed at surface and power transmitted to a haulage engine underground. Despite this, horses were still used in the branch tunnels and sledges on the coalfaces.

116a A horse being loaded onto the mine cage in the 1930s. *(Mrs A. N. Fielden)*

116b Two miners moving a sledge or 'dan' of coal along the coalface in the 1930s. Normally boys would have been used to haul the sledges using chains, but with the advent of motor-cycles giving easier access to alternative work, the boys left the mines and the older miners had to move the sledges themselves. *(Mrs A. N. Fielden)*

THE NORTH SHROPSHIRE OR OSWESTRY COALFIELD

This is really an extension of the Denbighshire Coalfield and, in Shropshire, covers an area of about sixteen square miles. Although this is only a small part of the coalfield, proportionally it has been quite productive. The last large mine, at Ifton, closed as recently as 1968. The coalfield contains seven important seams but the Two-Yard, Quaker, Powell and Main Coal seams have been the most important in recent years.

Ifton Mine, St Martins (SJ 321375). The mine was originally two separate concerns, Brynkinalt Colliery and Ifton Rhyn Colliery. Brynkinalt shafts were 165 yd deep and were sunk in the second half of the last century. The Ifton Rhyn shaft, 1½ miles east of Brynkinalt, is of a similar date but prior to deepening was only 135yd deep. During 1912-14 the Ifton shaft was deepened to 415yd and from this level two tunnels were constructed. These tunnels reached the coal seams in 1921 and 1923 respectively. A tunnel driven from Brynkinalt connected with Ifton shaft in 1923 and in 1928 Brynkinalt shafts ceased to be used for coalwinding. At this time the mine employed about 1,357 men. This was by far the biggest mine ever to operate in Shropshire. The mine was Nationalised in 1947 and shortly afterwards the nearby Black Park Mine was closed and its workings were incorporated in the former. Six workable coal seams were available spread vertically over 300ft of strata.

On Saturday 23 November 1968 the mine ceased production, due mainly to the geological difficulties being experienced. Immediately 413 of the 821 miners were declared redundant, others were pensioned off, a few remained to salvage the mine and some transferred to mines outside the county.

117 General view of Ifton Mine, about 1966. *(Shropshire Star)*

The South-West Shropshire Metal Mines

In comparison with other British orefields the Shropshire field is small and compact lying roughly within a radius of 6 miles around Shelve village, some 13 miles south-west of Shrewsbury. Despite its smallness the area has a long and interesting history; pigs of lead found locally bearing the stamp of Emperor Hadrian attest to Roman working while underground prospecting was still taking place in the 1960s. The chief minerals sought, roughly in chronological order of importance, have been the ores of lead, zinc, witherite, barytes, calcite together with a little fluorspar and copper.

The mineralisation occurs mainly in Ordovician strata, although some is found in the Pre-Cambrian rocks in the eastern part of the orefield. The distribution of the ores suggests some form of arched depth-zoning, the outer zone being barytes and witherite, below which lies lead, then zinc. The crown of the arch has been eroded away so that the deeper zone minerals now outcrop in the centre of the field west of the quartzitic Stiperstones ridge, while the barytes outcrops on both the eastern and western flanks of the orefield.

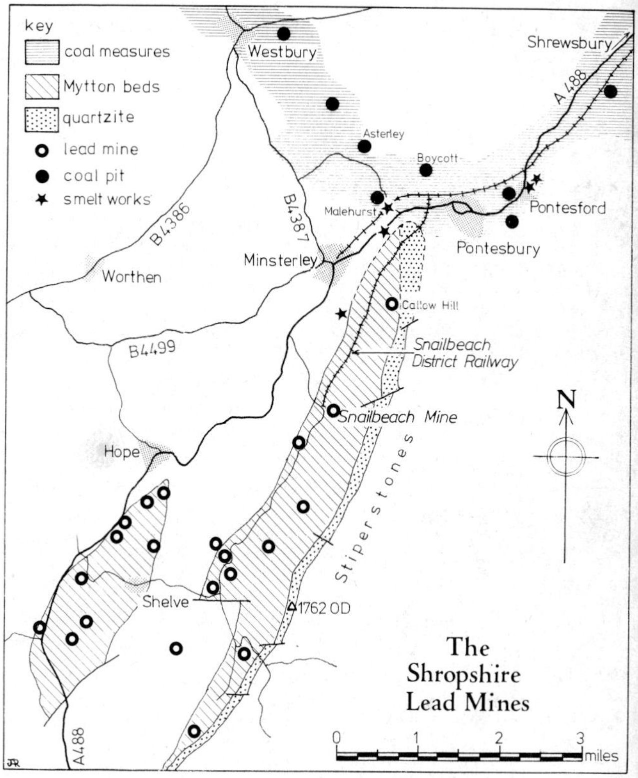

The most important mineral producer in the orefield was the Snailbeach Mine but other highly significant workings were the complexes based on Roman Gravels, Tankerville, The Bog and The Grits. In the present century the mines around Huglith have been foremost in the production of barytes, extraction ceasing as recently as 1948.

Snailbeach Mine (SJ 375022). Snailbeach Mine was certainly the richest lead mine in Shropshire (producing 131,900 tons of lead ore between 1845 and 1913, over half the ore produced in the area during that period, as well as other minerals) and is considered by many to have been one of the richest mines in Europe. A Roman pig of lead was found here in 1796 and there is documentary evidence of mining on the site in both the sixteenth and seventeenth centuries. Extensive working is known to have occurred in the eighteenth century and in 1782 the Snailbeach Mining Company was formed. By 1797 workings had reached 180yd depth, by 1872 462yd, by 1884 492yd and in 1899 552yd. In 1879 the mine employed 334 men. Lead mining ceased in 1913 but underground working continued on a limited scale for barytes until the 1950s.

118 *(above)* **Snailbeach Mine** about 1900, showing the mine before pumping ceased in 1911. The 'Snailbeach District Railway' narrow gauge locomotive on the left of the picture is near the foot of the incline which took wagons of coal up to the pumping engine. This railway, which was opened in 1877, can also be seen between the Black Tom shaft headframe and the white cottage to the right of the picture. The side-by-side double building in the centre of the picture is the loco shed. Left of this steam can be seen rising from the complex of buildings which included a winding engine of 1872 and an air compressor of 1881. In front of these buildings is the headframe of Old or George's Shaft. *(K. C. Lock)*

119 *(opposite, top)* **Snailbeach Mine Yard,** about 1910. The photograph shows the curved tramming level, by which loaded ore wagons were taken into the crusher house, after being drawn up the shaft. This view was taken from the west looking towards the position of the photographer in the previous picture. In that photograph the crusher house lies in front of the Old Shaft headframe. *(IGMT)*

120 *(opposite, bottom)* **Snailbeach Mine Reservoir and Pumping Engine,** about 1900. This photograph was taken from a position to the left of that used for Fig 118. The reservoir was constructed in 1872 to overcome the shortages caused by periods of drought. The mine pump on the hill to the left pumped normally only up to the drainage level and this discharged into the Hope Brook nearly 1,400yd away. In really dry weather however the pump could be used to raise water to the surface. In the foreground can be seen the rail tracks which brought mineral from the Lords Hill workings. *(K. C. Lock)*

SNAILBEACH MINES

122 **Old Shaft Headframe and Cage,** in 1967. This headframe, with single pulley, was used to wind barytes from above the 112yd Level after the pumps had stopped in January 1911 and the mine was flooded to adit level. The cage and headframe have deteriorated considerably since this photograph was taken, the frame itself having collapsed over the shaft.

122 **Old Shaft Headframe and Cage,** in 1967. This headframe, with single pulley, was used to wind barytes from above the 112yd Level after the pumps has stopped in January 1911 and the mine was flooded to adit level. The cage and headframe have deteriorated considerably since this photograph was taken, the frame itself having collapsed over the shaft.

123 **The Blacksmith's Shop, Snailbeach,** about 1900. The long drill steels shown on the truck have been brought from the mine for sharpening. These long steels were used with the compressed-air drilling machines which had been introduced to the mine in 1881. By 1882 Snailbeach had four drilling machines in use. Such innovations were not without dangers and in 1883 the 'foreman of boring machinery' at Snailbeach was killed and two miners seriously injured when the blows of their drill initiated a charge of dynamite. *(K. C. Lock)*

124 **Ore Dressing Plant at Snailbeach,** 1960. A simple jig and spiral classifier of unknown date standing close to Black Tom Shaft. It is believed that this timber-constructed plant was used for the preparation of barytes up to the 1920s.

125 *(opposite, top)* **Ore dressing plant, Snailbeach,** 1960. This plant stands outside Perkins' Level and was used for the preparation of barytes up to the mid-1950s. A large stockpile of barytes and several kibbles of lead ore can be seen in the foreground. These had been separated from the run of mine material by a simple jigging and screening operation.

126 *(opposite, bottom)* **Spar-separating plant, Snailbeach,** 1960. The dressing of spar (quartz and calcite chips) from the waste heaps for use in the building industry for such purposes as pebble dashing continues to the present day. The plant shown was used in the post-war period and consists of a receiving hopper, elevating conveyor, rotary screen or trommel and bins. The plant was scrapped shortly after the photograph was taken.

127 **The Barytes Vein in Perkin's Level,** 1970. These workings under Lords Hill are quite extensive, and as can be seen the vein is about 55 degrees and up to 7ft wide in places. Pillars of white barytes left in to support the hanging wall can also be clearly seen. The vein was worked in this part of the mine as recently as the 1940s. *(Shrewsbury Chronicle)*

128 **A Blow-George or Ventilating Fan,** 1967. This small ventilating fan consisted of a number of blades in a close-fitting chamber and an outlet (bottom right) through which the air was forced down a tube and into the mine. This fan had been situated on a timber plank over an open shaft above the Perkin's Level workings on Lords Hill, Snailbeach. It was driven by a belt from a farm tractor or other mobile power source.

129 **Smelter Flue, Snailbeach Smelt house,** 1967. This underground condensing chamber was at the confluence of the flues from the five hearths at the 'New' Snailbeach Smelt House. This smelt house was opened in 1862 and closed in 1895, although one hearth was in use for a short period in June 1897. The opening ceremony on 11 July 1862 consisted of a procession by 400 employees, followed by a dinner with ale set out on three rows of tables 145ft long placed inside the new smelthouse. A length of flat wire rope from the mine can be seen on the left of the photograph. This had been used as a safety fence.

Snailbeach District Railway. The Snailbeach District Railway was of 2ft 4in gauge and was opened to traffic in 1877. The railway was just over three miles in length and connected the Snailbeach Mine to Pontesbury siding on the GW-LNW branch line from Hanwood Junction to Minsterley. Steam locomotion ceased in 1946 but the wagons continued to be used to carry stone from a quarry adjacent to the line, hauled by an ordinary farm tractor, until 1959.

130 and 131 **Snailbeach District Railway Locomotive No 2,** about 1935. This was one of six steam locomotives used on the railway. It was an 0-4-2T built in 1902 by Kerr Stuart & Co Ltd of Stoke on Trent. After a chequered history it was purchased by the line's owner, Colonel Stephens, in 1922. Although at one time named *Skylark,* on arrival on the SDR it became simply 'No 2'. Locos Nos 2, 3 and 4 worked all the traffic from 1923 up to the end of steam working in 1946. Each was used in turn for spells of 2-3 weeks which meant, unfortunately, that eventually all became unserviceable at about the same time. The locos were cut up on site by Thos W. Ward Ltd in May 1950. *(K. C. Lock)*

132 **Pontesbury Transhipment Sidings,** 1961 (SJ 393062). The hopper wagons of the Snailbeach District Railway are on the stout oak pier-supported siding standing above the standard gauge metals. To tranship from SDR to main line wagons was merely a matter of opening the trap door in the hopper — but for transhipment in the opposite direction heavy shovelling was necessary. The rails and wagons were removed shortly after this photograph was taken; some of the rails went to railway preservation societies but unfortunately the wagons were all scrapped.

133 *(opposite, top)* **The Loco Shed at Snailbeach,** about 1950. The shed lies between the mine buildings and the spoilheaps. Black Tom headframe can be seen in the trees to the left of the shed. Loco No 3 (a Baldwin rebuilt by Bagnalls in 1918) stands outside the shed and was scrapped shortly after the photograph was taken. *(J. D. How/K. C. Lock)*

134 *(opposite, bottom)* The Loco Shed as it appeared in 1961.

135 and 136 *(opposite)* **Reworking the Spoil Tips at Snailbeach,** about 1930. The spoil tips have been worked since about 1900 for 'spar' to be used in the building industry. For many years this was done by the Halvans Company (halvans is a Cornish term for mine waste) and later the Gravel Trading Co. The photographs show the dressing plant in use about 1930 with a jig washer in the foreground, and the junction of the 'spar' siding with the Snailbeach District Railway. Spar working continues on the spoil tips at the present time. *(K. C. Lock)*

137 **The Halvans Company Engine House,** in 1960. This company worked the spoil tips of the mine for quartz, calcite and barytes from about 1900 to the 1920s. In addition the company also mined barytes from the Black Tom Shaft. This building housed the steam engine used to drive the ore-dressing plant, a small part of which can be seen in the foreground.

138 and 139 **New Central Mine,** 1864 (SJ 369016). These photographs taken from a shares prospectus are the oldest known photographs of a mine in this field. The mine was worked unsuccessfully from 1864 to 1868 by the Central Snailbeach Company. During this time it reached a depth of 164yd. A truncated square chimney and a boiler house complete with Lancashire boiler, now used for storing water, remain on this site.

The pictures show shaft sinking in progress with the spoiltips being very newly-formed. One of the pictures shows a timber cover, on what appears to be an air shaft, in the foreground. *(K. C. Lock)*

140 and 141 **Burgam Lead Mine** 1960 (SO 358996). The pictures show the mine during prospecting for lead by Mr J. L. Burden under the name of Jackfield Mines Ltd. This old mine was re-opened in 1959 and a couple of short adits driven into the hillside by two local miners using very primitive equipment. The classical prospector's hut (complete with tin bath), a timbered adit mouth with wheelbarrows and samples are shown. This prospecting had ceased by 1963. The earlier workings here had been in the Big Spar Lode, which was mainly a barytes vein with stringers of galena and blende. Extraction of these minerals is recorded from the middle of the nineteenth century to 1940, but during this century operations have been very intermittent.

Tankerville Mine, about 1870 (SO 355995). Considerable work was carried out at this mine after 1870. Watson's Shaft was sunk to 190 fathoms by 1874 and after further sinking eventually became the deepest shaft in the orefield. Originally there was a steam engine underground with a chimney to surface of 1,140ft length. Later this was replaced by a steam engine at the surface and a Cornish-type beam engine was also added. The mine finally closed in 1894 and the engines and equipment were sold by auction in 1902.

142 *(opposite top)* The miners line up for the photographer. Behind them to the right the headframe over the Old Ovenpipe Shaft and to the left that of Watson's Shaft. *(K. C. Lock)*

143 *(opposite, bottom)* A close-up of Watson's Shaft headframe and the mine yard is shown in the second picture. Both photographs were taken before the present large engine house was built. *(K. C. Lock)*

144 **Roundhill Mine,** about 1910 (SO 351996). Despite the caption this is not Tankerville Mine. Roundhill Mine was at work, mainly in a limited way, throughout the second half of the nineteenth century. In the early years it was mainly a lead mine, but later became a barytes producer. By 1900 the mine had closed, but was re-opened again in 1906 and closed in 1913. It was re-opened again in the early 1920s. The steam engine here was built at the Perran Foundry, Falmouth in 1874. *(K. C. Lock)*

Bog Mine (SO 356978). This is a very old mining site and the Bog Mine is recorded as early as 1739. By 1777 the mine had a Boulton and Watt steam pumping engine with a 30in diameter cylinder. The mine had a chequered history and by the end of the nineteenth century employed only nine men producing lead and zinc ores and barytes.

145 Bog mine about 1910. The tall newish building in the centre contained 'National' gas engines driving Crompton Parkinson alternators to produce electricity. The gas for the engine was made on site from coal and anthracite brought in from Wales. From about 1910 operations expanded considerably and under The Shropshire Mines Ltd over a hundred men were employed. A new shaft called Ramsden Shaft was sunk and the first electric winder in Shropshire installed on this shaft in 1915. This was unsuccessful and was replaced by a steam winding engine. Much of the work in this period was done around Buntings Shaft (the headframe of which can be seen above the buildings) and the principal mineral produced was barytes. The mine closed about 1924 although work continued for a few more years at the nearby associated Rock Mine. *(K. C. Lock)*

146 Bunting Shaft headframe 1960. This shaft has been used for both pumping and winding. It is connected to the Boat (or Drainage) Level at about 104yd depth. Shortly after the photograph was taken the headframe was dismantled and the area partially cleared.

147 **Roman Gravels Mine,** about 1890 (SO 334998). This site is reputed to have been mined for lead by the Romans and workings in this vicinity are certainly of great antiquity. Despite this, the mine reached its peak toward the end of the nineteenth century and by 1890 still employed two hundred men, fifty more than Snailbeach Mine at that time. To the right of the picture can be seen Old Engine Shaft head-frame. This shaft has been used for both pumping and winding ore. The mineral passed over the triple-decked bridge to the mine's extensive dressing floors on the west side of the main road. The buildings in the foreground were the carpenters' and smiths' shops. Many of the miners at the time can be seen lined up on the upper deck of the bridge. The mine closed about 1900. *(Shropshire Magazine)*

Wotherton Mine (SJ 277030). This mine produced barytes from 1865 to November 1911, during which time over 125,000 tons was extracted. In early years pumping and winding was carried out at Old Engine Shaft, although in later years winding was done at New Engine Shaft. Throughout its life the mine employed between thirty and forty men.

Since the mine closed the engine house at Old Engine Shaft, of shaped red sandstone blocks, has been converted into a dwelling. A segment of a circle is cut into the stonework at the lever-wall end to take the flywheel, a feature seen on old photographs of steam whims in Cornwall.

148 *(below)* The engine house at Old Engine Shaft, 1967, showing the space left in the south wall for the flywheel.

149 *(opposite, top)* A general view of the mine about 1885. Large heaps of barytes can be seen, together with one form of transport used from this rather isolated mine although most of the barytes was hauled to Hanwood Mill using road locomotives.

150 *(opposite, bottom)* Wotherton miners about 1905. Some of the miners are in their working clothes (notice the wide-brimmed hard miners' hats), while others obviously came in for the occasion. *(T. Gwilliam)*

151 and 152 **The Miners' Arms Hotel, Minsterley,** about 1915 (SJ 375051). As in every mining area there are many inns, several of which, like the one shown, were quite imposing. This particular inn stands at a crossroads beside a road bridge. The bridge opened in 1910. With the closure of the mines the name of this inn was changed and it is now called simply the Bridge Inn. (Hidden from view and to the left of the photograph there is a very fine Victorian gentlemen's cast-iron urinal).

Another interesting inn (also called The Miners' Arms) closed during the 1960s. This was at The Bog, a hamlet which is now virtually deserted. *(K. C. Lock)*

153 **Road Locomotive,** 1917. This engine was used to haul ore and barytes from the mines of The Shropshire Lead Mines Ltd, particularly the Bog Mine, to the station at Minsterley. The engine was a Foster-Wellington tractor weighing 6 tons. It was of 4 nominal horsepower (approximately 35 brake horsepower) and carried extra water tanks. Although one of the smaller tractors in use in the area, this one was capable of hauling loads of six to eight tons.

From 1917 to 1924 the work of carrying ore and barytes from the mines was done by a 5½ mile long aerial ropeway. This was constructed with labour supplied by German prisoners of war, housed locally in a railway workshop. The rope buckets also took the anthracite and coal up to the mine for conversion to gas for the gas engines. *(K. C. Lock)*

154 **Aerial Ropeway** near Malehurst, Minsterley, about 1930. There were several aerial ropeways in the orefield in the early part of the present century and the actual location of the one shown is not known. It was probably somewhere near Malehurst Mill on the 3½ mile ropeway constructed about 1922 to carry barytes from the Huglith Mine. A timber trestle is shown but steel trestles were used near public highways. The ropeway was dismantled in 1949. *(K. C. Lock)*

155 **Huglith Mine,** about 1930 (SJ 405015). This was a highly productive barytes mine which commenced production in 1910 and reached a peak, exceeding 20,000 tons per annum, in the mid 1930s. The mine finally closed in 1945 having produced over 290,000 tons. At most the mine employed sixty-five men and a group of these, with Tom Hocking to the front, are shown in the photograph. Mr Hocking later became the mine manager but does not seem to have been accorded the courtesy title of captain, normally given to the manager of Shropshire's metal mines.

156 Drilling a shot-hole in the barytes vein at Huglith Mine in the late 1930s. A compressed-air drill is being used, held secure by a steel threaded prop. A typical set of hardrock miners' tools can be seen to the left: long and short drills, hammer, pick and shovel. At this time candles were still being used for illumination and these were stuck to the front of the miner's hard hat by a gob of clay; the runs of fat can be seen on the miner's hat on the right.

157 and 158 **The Sallies Mine,** 1946 (SJ 396001). These photographs were taken during a visit of executives from the Laporte Chemical Company. The underground photographs show overhand stoping for barytes in progress. Compressed-air drills are in use to bore the shotholes and there is a device present for reducing the amount of dust produced. During this time the mines became quite notorious for the incidence of silicosis or 'dust disease' occurring there and the miners were encouraged to wear masks. This mine was opened about 1937 and closed in 1948. During the later years nearly forty men were employed at the mine. *(F. Lakelin and K. C. Lock from originals by the West Midland Photo Company)*

159 **Hanwood Barytes Mill,** about 1920 (SJ 441093). The white barytes from practically all the mines to the west of the Stiperstones except for Weston and Calcot Mines was dealt with at this mill. There was a corn mill on the site before 1800 and by 1810 it had become a Flax, Yarn and Thread Mill. By 1886 it had become a barytes mill owned by the Wotherton Barytes and Lead Mining Co. In 1915 the processing consisted of crushing, bleaching (using sulphuric acid), washing and milling using a pair of rolls. In 1919 the mill was operated by The Shropshire Mines Ltd whose stated policy was 'to produce the best quality of ground barytes which science can devise'. The mill closed in 1922.

Other barytes mills included those at Cliffdale, Sutton Mill (Shrewsbury) and Malehurst. The latter worked from 1922 to 1948 and when this opened all the other mills closed. Since 1956 Malehurst Mill has been used for the preparation of animal foodstuffs.

160 **Cornish Engine House** believed to be at Westcott Copper Mine (SJ 402012). The photograph was taken in 1918. No other early photograph of a Shropshire copper mine is known. The much reduced ruins of these structures can still be seen. *(J. Price and K. C. Lock)*

Selected bibliography

Adams, D. R. (ed), 1962, *Survey of the South Shropshire Lead Mining Area.* First Supplement, 1968; Second Supplement by Brown, I. J. and Davies, T. J., 1972. All published by the Shropshire Mining Club.

Brook, F. and Allbutt M., 1973, *The Shropshire Lead Mines.* Moorland Publishing Co.

Brown, I. J. 1965, 'The Mineral Wealth of Coalbrookdale', *Bulletin of the Peak District Mines Historical Society,* Vol 2, Pts 5 and 6.

Brown, I. J., 1966, 'A history of limestone mining in Shropshire', *Shropshire Mining Club,* Dec 1966.

Brown, I. J., 1968, 'Coal Production in South Shropshire', *Shropshire Newsletter* (Shrops Arch Society), No 34, Jan 1968.

Brown, I. J., 1968, *The Coalbrookdale Coalfield Catalogue of Mines and Mining Bibliography,* Shropshire County Library.

Brown, I. J., 1973, *Mine accidents in the South Shropshire Metalliferous Mining Area,* 1875—present. (Shropshire Mining Club.)

Brown, I. J., 1975, *Mineral Working and Land Reclamation in the Coalbrookdale Coalfield,* Ph.D. thesis, Leicester University.

Davies, T. J., 1969, *The Engine Houses of the Mines of South Shropshire* (Shropshire Mining Club).

Dines, H. G., 1958, *Bulletin of the Geological Survey,* No 14, HMSO

Clayton Jones, H., 1960, 'Shropshire Mining Outpost — the three pits of St Martins', *Shropshire Magazine,* March 1960.

Lock, K. C., 1970, 'When Shrewsbury had its own coalfield', *Shropshire Magazine,* January 1970.

Pocock, R. W., Whitehead, T. H., et al, 1938, *Memoirs of the Geological Survey of Great Britain,* 'Shrewsbury District'.

Pocock, R. W. and Whitehead, T. H., 1948, *British Regional Geology,* 'The Welsh Border Land' HMSO

Price, G. L. A., 1968, 'Coal Mining in the Clee Hills of South Shropshire', *Trans Birmingham Enterprise Club,* Vol 2, March 1968.

Tonks, E. S., 1950, *The Snailbeach District Railway* (2nd edition 1974), The Industrial Railway Society.

Trinder, B. S., 1973, *The Industrial Revolution in Shropshire,* Phillimore & Co.

Whitehead, T. H., and Pocock, R. W., 1947, *Memoirs of the Geological Survey of Great Britain,* 'Dudley and Bridgnorth'.

Whitehead, T. H., and Robertson, T., et al, 1928, *Memoirs of the Geological Survey of Great Britain,* 'Wolverhampton and Oakengates'.

Useful sources of information on Shropshire mines are the records held by the County Archivist, Shirehall, Shrewsbury, In particular, for the metalliferous mining field, the collection deposited by Mr. K. C. Lock contains much interesting material. Another collection presented by the Shropshire Mining Club also includes some historic documents. In addition there are other collections available of a more specialised nature.

General Index

accidents, 32, 66, 86, 87
aerial ropeway, 73, 105
air compressor, 84

barytes mills, 105, 108
barytes working, 83, 86, 87, 89, 95, 97, 99, 100, 102, 106, 108
bell-pit workings, 78, 79
bitumen, 13, 26
blast furnaces, 17, 31, 44
Blists Hill Museum, 5, 24, 26, 47
Boat Level, 100
Brown Clee Coalfield, 78

calcite working, 83, 89, 95
canals, 26, 31, 63, 81
candles for illumination, 54, 68, 106
Catherton Common, 79
chartermasters, 23, 32, 34
children in mining, *frontispiece,* 81
clay working, 8, 24, 30, 33, 50, 51, 54
Clee Hill Coalfields, 78
Cliffdale Mill, 108
Coalbrookdale Coalfield, 8-67
Coalbrookdale Company, 15, 60
coal cutting machines, 21, 34, 48, 54, 68, 73
coal distillation, 44
coal wharves, 67, 71
Coke, 26
compressed-air picks, 76
concrete headframe, 72
copper working, 6, 83, 108
crushing, 84, 108

Denbighshire Coalfield, 82
donkeys underground, 56
drainage, 24, 33, 41, 42, 61, 84, 86, 100
drilling machines, 54, 68, 87, 106
Dryton Coalfield, 80
dust suppression and disease, 107

electric power generation, 81, 100
electric winding engines, 19, 34, 39, 47, 48, 54, 76, 81, 100
explosions, 66, 87
explosives, 60, 87

fire bucket ventilation, 54, 55
flat winding ropes, 34, 39, 40, 42, 90
Forest of Wyre Coalfield, 68-77
fossils, 12

gas engines, 100, 105

handwinch, 31, 45, 54, 55
Hanwood and Moat Hall Collieries Ltd, 81
Hanwood Coalfields, 80
Hanwood Mill, 102, 108
Highley Mining Company, 68-75
horse gins, 10, 28, 50, 59, 62
horses, 19, 24, 28, 81
housing, 62, 80, 102

inclined shaft, 86
Ironbridge Gorge Museum Trust, 5, 61
ironstone working, 9, 19, 23, 32, 33, 41, 78

jigs, 87, 89, 95

kibbles, 89

ladderway, 86
lead pigs, 83
lead working, 83, 99, 101
Leebotwood Coalfield, 80
Lilleshall Company, 33-44, 60, 61, 62
limekilns, 11, 56
limestone working, 6, 8, 33, 56, 78, 80
locomotives underground, 21, 34, 74

Madeley Court Company, 31, 32
Madeley Wood Company, 17-31, 61
Malehurst Mill, 105, 108
methane drainage, 39, 40
miners' co-operative, 76
miners' inns, 104
Mole Mining Ltd, 76
Moody, W. L., 75

North Shropshire Coalfield, 82

ore dressing plant, 87, 89, 95
Oswestry Coalfield, *frontispiece,* 82

pillar and stall working, 48, 54, 56
Pontesbury Transhipment Siding, 92
Prestwich, J., 56
Priorslee Village, 62
prospecting, 13, 26, 83, 97
pyrites working, 54

quartz working, 89, 95

railways, 71, 84, 91, 92, 94
Randall, J., 59
reservoir, 84
Reynolds, W., 63
river trows, 67
road locomotives, 102, 105

roof support, 50, 72, 73, 89
rope haulage underground, 20, 37
screening, 41, 48, 89
shaftsinking, 31, 33, 34, 72, 82, 96, 99
Shrewsbury Coalfield, 80
Shropshire Mines Ltd, 100, 105
Shropshire Mining Club, 26, 54, 56, 62, 108, 109
sledges, 65, 81
smelters, 80, 90
Snailbeach District Railway, 84, 91, 92, 94
Snailbeach Mining Company, 83
spar working, 89, 95
steam locomotives, 91
steam pumping engines, 9, 16, 28, 30, 32, 33, 35, 36, 39, 41, 61, 66, 80, 84, 100, 102, 108
steam winding engines, 9, 15, 16, 17, 19, 23, 24, 28, 30, 32, 33, 34, 35, 36, 39, 40, 42, 43, 48, 51, 53, 68, 84, 96, 99, 100, 102
stoping, 89, 107
strikes, 45, 64, 65

Tar Tunnel, 11, 26, 63
Titterstone Clee Coalfield, 78
tramways, 28, 51, 52, 60, 62, 63, 74, 77, 84
tub boats, 31, 42
tunnels, 31, 63, 82

vein workings, 89, 107
ventilating fans, 23, 33, 34, 39, 42, 71, 90
ventilating furnaces, 10, 19, 34

winding cages and boxes, 24, 33, 40, 52, 55, 74, 86
winding chain, 15, 33, 56, 57, 60
witherite working, 83
women mineworkers, 41
workmen's carriages underground, 75

zinc working, 83, 100

Mines Index

Alveley, 68, 72-75

Bayton, 75
Billingsley, 68
Black Park, 82
Black Tom, 84, 87, 92
Blists Hill, 24, 26, 30
Bog, 100, 105
Brandlee, 48
Brynkinalt, 82
Burgam, 97

Calcot, 108
Cats Head, 61
Cruck Meole, 81

Deep (Guest's), 51, 52, 53, 59

Foxholes, 13, 14, 65
Freehold, 42

Grange, 33, 39, 40
Granville, 33, 34-39
Gravel Leasows, 15

Hadley, 9
Halesfield, 20,23
Hanwood, 81
Highley, 68, 69, 71
Hollygrove, 50
Huglith, 83, 105, 106
Hunthouse, 76

Ifton, 82
Ifton Rhyn, 82
Ivy, 45

Kemberton (Madeley Wood), 17, 18, 19

Lane, 28
Lightmoor, 59
Lincoln Hill, 56, 57
Lloyds, 61
Lords Hill, 84, 89

Madeley Court, 31, 32
Madeley Wood, 28
Milburgh, 24

Nabb, 45
New Central, 96

Old Park (Harris), 47

Perkins, 89, 90
Pontesford, 80
Preesgweene, *frontispiece*
Pudley Hill, 61

Rats Pit, 16
Rock, 54, 55
Roman Gravels, 101
Roundhill, 99

Sallies, 107
Shawfield, 30

Shortwoods, 48
Snailbeach, 83-95
Stafford, 33

Tankerville, 99

Waxhill Barracks, 41
Westcott, 108
Weston, 108
Wombridge, 61
Woodhouse, 42
Wotherton, 102